# Centrifuge Modelling for Civil Engineers

# Centrifuge Modelling for Civil Engineers

Editor

**Devendra Damale**

**scitus**
academics

**Centrifuge Modelling for Civil Engineers**

Edited by **Devendra Damale**

Printed in 2017

ISBN: 978-1-68117-014-5

Library of Congress Control Number: 2015931521

© 2016 by
SCITUS Academics LLC,
616, Corporate Way, Suite 2, 4766,
Valley Cottage, NY 10989

www.scitusacademics.com

Printed in United States of America on Acid Free Paper ∞

# Contents

# Preface

Centrifuge modelling involves testing of small scale models in the enhanced gravity field of a geotechnical Method centrifuge. This technique is particularly useful in testing materials such as soils which exhibit non-linear stress-strain behaviour and can suffer significant plastic strains. Centrifuge modelling is proven to be particularly effective in determining the failure mechanisms for a wide variety of geotechnical problems. The scale model is typically constructed in the laboratory and then loaded onto the end of the centrifuge, which is typically between 0.2 and 10 metres (0.7 and 32.8 ft) in radius. The purpose of spinning the models on the centrifuge is to increase the g-forces on the model so that stresses in the model are equal to stresses in the prototype. A geotechnical centrifuge is used to test models of geotechnical problems such as the strength, stiffness and capacity of foundations for bridges and buildings, settlement of embankments, stability of slopes, earth retaining structures, tunnel stability and seawalls. Other applications include explosive cratering, contaminant migration in ground water, frost heave and sea ice.

This book summaries a generalized design process engaged for civil engineering projects. This book is a reference tool for graduate students, researchers, and practicing civil engineers involved with geotechnical issues.

**Editor**

# Chapter 1

# Centrifuge Modeling of Rocking-Isolated Inelastic Rc Bridge Piers

Marianna Loli[1], Jonathan A, Knappett[2], Michael J. Brown[2], Ioannis Anastasopoulos[2], and George Gazetas[1]

[1]School of Civil Engineering, National Technical University of Athens, Greece
[2]Division of Civil Engineering, University of Dundee, UK

## ABSTRACT

Experimental proof is provided of an unconventional seismic design concept, which is based on deliberately under designing shallow foundations to promote intense rocking oscillations and thereby to dramatically improve the seismic resilience of structures. Termed rocking isolation, this new seismic design philosophy is investigated through a series of dynamic centrifuge experiments on properly scaled models of a modern reinforced concrete (RC) bridge pier. The experimental method reproduces the nonlinear and inelastic response of both the soil-footing interface and the structure. To this end, a novel scale model RC (1:50 scale) that simulates reasonably well the elastic response and the failure of prototype RC elements is utilized, along with realistic representation of the soil behavior in a geotechnical centrifuge. A variety of seismic ground motions are considered as excitations. They result in consistent demonstrably beneficial performance of the rocking-isolated pier in comparison with the one designed conventionally. Seismic demand is reduced in terms of both inertial load and deck drift. Furthermore, foundation uplifting has a self-centering potential, whereas soil yielding is shown to provide a particularly effective energy dissipation mechanism, exhibiting significant resistance to cumulative damage. Thanks to such mechanisms, the rocking pier survived, with no signs of structural distress, a deleterious sequence of seismic motions that caused collapse of the conventionally designed pier. © 2014 The Authors Earthquake Engineering & Structural Dynamics Published by John Wiley & Sons Ltd.

# BACKGROUND AND OBJECTIVES

Capacity design, which forms the cornerstone of modern seismic design, aims at controlling seismic damage by strategically directing inelastic deformation to structural components, which are less important to the overall system stability [1]. Although this design approach is enforced or at least encouraged by modern seismic codes, it is conventionally limited to the superstructure. The foundation system is contrastingly treated conservatively. Specifically, the current foundation design leads to a strong unyielding foundation–soil system by adopting over strength factors to ensure that their ultimate capacity is reliably greater than the largest moment to be transmitted by the pier column.

Field evidence suggests that although new structures, complying with this capacity design rationale, are generally safer than the older ones, they remain vulnerable to very strong shaking. In fact, an alarmingly large number of modern structures have suffered intense damage leading to partial or total failure in recent earthquakes [e.g., [2, 3]] signaling the need to rethink the effectiveness of the aforementioned design practice. In response, a number of studies have explored the possibilities and constraints of an alternative design concept: allowing the development of 'plastic hinging' in the soil or at the soil–foundation interface, so as to reduce the possibility of damage to the elements of the structure.

Focusing on surface foundations, where nonlinearity manifests itself through uplifting and/or soil yielding, a 'reversal' of the current capacity design principle is proposed: the foundation is intentionally under designed under the seismic actions compared with the supported column to promote rocking response and accumulation of plastic deformation at the soil–foundation interface. Supporting evidence for this new approach has been provided by the following theoretical and empirical findings:

- Several theoretical and numerical studies on the rocking response of rigid blocks and elastic single-degree-of-freedom (SDOF) oscillators [e.g., [4-7]] provide compelling evidence that uplifting drastically reduces the inertial load transmitted into the oscillating structure.

- Because of the transient and kinematic nature of seismic loading, rocking response does not lead to overturning even in the case of very slender structures [8-14], except in rather extreme cases of little practical concern.

- Referred to as rocking isolation, allowing for foundation uplift has been proposed, and in a few exceptional cases employed in practice, as a means of seismic isolation [15-17].

- Even in the case of relatively heavily loaded footings or footings on soft soils, when rocking is accompanied with yielding of the supporting soil (and possibly momentary mobilization of bearing capacity failure mechanisms), substantial energy is dissipated in the foundation providing increased safety margins against overturning owing to the inherently self-centering characteristics and the ductile nature of rocking on compliant soil [e.g., [18-23]].

- Most importantly, a number of studies have recently investigated the scheme of rocking isolation, with emphasis on its effects on structures, which consistently point to a beneficial role of nonlinear foundation behavior for the overall system performance. Previous studies include the numerical and experimental work of

[24-26] in the domain of framed building structures, as well as those of [27-30] in the domain of bridges.

- A variety of modern numerical tools have been developed enabling comprehensive modeling of nonlinear rocking response [31-37], alleviating to some degree the skepticism regarding the uncertainties traditionally associated with prediction of the performance of rocking foundations for use in design.

On the basis of the exploratory work of [27], this study seeks to provide experimental verification of their numerical findings suggesting that although a conventionally designed reinforced concrete (RC) pier on an adequately large shallow foundation would suffer structural failure of the RC column and collapse in an earthquake sufficiently exceeding its design limits, rocking motion of an alternative under designed foundation would allow the same pier to survive even extreme shaking scenarios. To this end, it was necessary to realistically model the various attributes of nonlinear response of both the structural element (RC column) and the soil-foundation interface, therefore necessitating the use of the following:

- A new scale model reinforced concrete [38] capable of replicating stiffness, strength, failure mode, and post-failure response of the bridge pier.
- The University of Dundee (UOD) centrifuge earthquake simulator (EQS) to accurately replicate nonlinear soil behavior and provide repeatable replication of historically recorded earthquake motions.

A series of centrifuge tests were performed to investigate and compare the performance of the two RC model bridge piers, having the same structural section in each case, but each representing one of the two considered design approaches, namely, a conventional design and a rocking isolation design. Presented in this paper are the results from four of these tests involving a variety of shaking scenarios using real historical ground motions? Previous studies have simulated similar problems in the centrifuge by introducing reduced structural model cross sections (mechanical fuses) to control the locations of inelastic deformation and strength within structures usually made of aluminum [e.g., [29, 39]]. Despite the valuable insights of such testing, there are a number of unavoidable limitations:

- The location of any plastic damage must be defined a priori and any moment redistribution within the section is therefore suppressed.
- The axial load–moment capacity interaction behavior for aluminum is not of the same shape as for an RC section (in which axial load initially increases the moment capacity); therefore, any axial load redistribution within the section will not be captured correctly.
- The response of a fuse (whether in the form of notches or of artificial hinges) will not show the degradation in the moment–curvature behavior typical of reinforced concrete sections under cyclic loading.

The tests presented herein therefore attempt to model the inelastic response of RC elements in the centrifuge more realistically, namely, by using a new scaled model reinforced concrete consisting of a geometrically scaled steel reinforcing cage embedded within a cementations material, while simulating the entire soil–foundation–structure system as a whole. In this way, the correct foundation behavior can be modeled simultaneously with a structure having fidelity of response closer to that of a larger scale structural element test than has been achievable to date. Hence, the paper pursues two additional objectives: (i) presentation of the design and construction of realistically scaled RC model piers for use in centrifuge experiments and (ii) a report of 1–g calibration testing to verify the bending behavior of these piers. In the following, properties and results are given at prototype scale, unless otherwise stated.

# DESIGN OF THE PIER–FOUNDATION SYSTEMS

Figure 1 illustrates the conceptual prototype of a modern bridge pier designed in accordance with Eurocode specifications for RC structures and seismic actions [40, 41]. The deck is a cast-in-place concrete box girder with a total dead load $q_1 = 200$ kN/m, free to rotate oscillating in the transverse direction. Thus, a 10.75 m tall (to the center of mass of the deck, including the footing) cantilever is considered carrying the concentrated mass of the supported deck ($m^{deck} = 300$ Mg) and comprising a RC column with a cross section of 1.5 m × 1.5 m (cross-sectional area $A_c = 2.25$ m2).

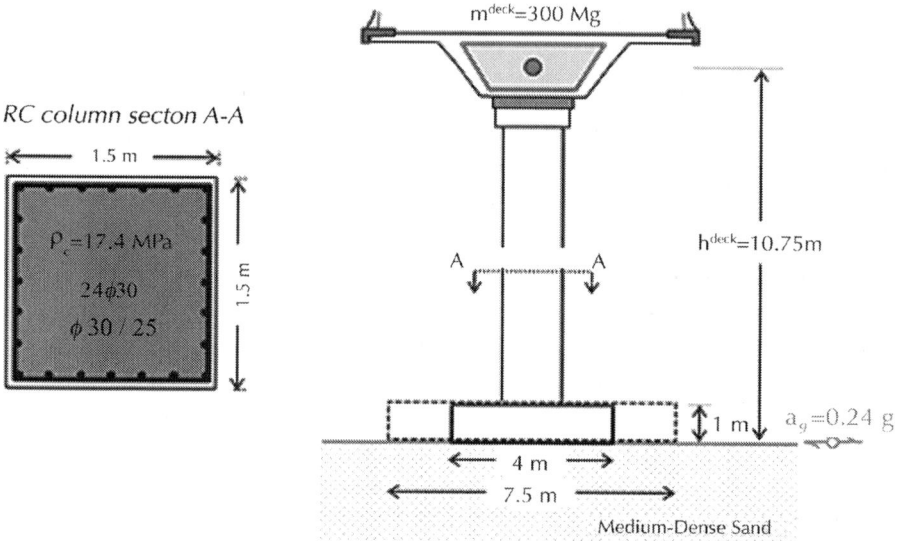

*RC column secton A-A*

Figure 1: Schematic definition of the studied problem.

The RC column was designed to withstand static loads and seismic actions due to a design ground acceleration $\alpha_g = 0.24$ g. Using the EC8 specified acceleration response spectrum for a Type C soil profile and assuming ductile behavior (q = 3), the design spectral acceleration may be calculated as $S_A = 0.23$ g. Considering the availability of typical model reinforcement sizes and properties (discussed in the following section) as well as reasonable geometries to allow ease of construction, it was deduced that uniformly spaced longitudinal reinforcement of 24 bars of $d_{bl} = 30$-mm diameter combined with shear links of the same diameter spaced at 250 mm was required to effectively carry design loads, using a concrete of compressive strength $f_c = 17.4$ MPa (cylinder strength). Table 1 summarizes calculations regarding design loads (with reference to the base of the column) and verification of adequate reinforcement in shearing and bending (moment capacity and ductility). It should be noted that bending response was predicted through cross-sectional analysis employing the USC_RC software [42].

**Table 1:** RC column load calculations and design assessment with respect to the Eurocode

| Loading and Resistance of RC section | | |
|---|---|---|
| RC, reinforced concrete. | | |
| Axial load | N:MN | 3.426 |
| Shear load | V:MN | 0.677 |
| Moment load | M:MNm | 6.600 |
| Normalized axial load | $n$ (=$N/A_c f_c$) | 0.088 |
| Normalized shear load | $v$ (=$V/A_c f_c$) | 0.017 |
| Normalized moment load | $m$ (=$M/bA_c f_c$) | 0.112 |
| Moment capacity | $M_u$:MNm | 6.816 |
| Factor of safety | FS | 1.1 |
| Curvature ductility | $\mu_\varphi$ | 18 |
| Displacement ductility | $\mu_\Delta$ | 6.5 |
| Capacity design shear | $V_{Ed}$:MN | 1.022 |
| Shear reinf. yield shear force | $V_{Rd,s}$:MN | 3.415 |
| Maximum member shear force | $V_{Rd,max}$:MN | 10.257 |
| Shear resistance | $V_{Rd}$:MN | 3.415 |
| Factor of Safety | FS | 3.3 |

The pier is founded on a shallow, 10-m thick, layer of medium density sand with a square (B × B) footing. Two different footing dimensions are studied, representing the two design alternatives. Table 2 summarizes the two foundation designs listing: static and seismic loads (QE, ME), design seismic actions ($QE_d$, $ME_d$), and ultimate shear and moment capacities ($Q_u$, $M_u$) and the factors of safety for static vertical loads (FSV) and seismic loads (FSE). It should be noted that, apart from the small difference in the total vertical load $N_{tot}$ (resulting from the different foundation sizes), the two piers are subjected to the same design loads (as they are identical in geometry) and to the same design ground acceleration $\alpha_g$. $QE_d$ and $ME_d$ are calculated according to the capacity requirements for the foundation overstrength, which calls for seismic design actions on the foundation be substantially magnified (by as much as 40% in the case of a cantilever structure) in comparison with the actual seismic loads at the column base. Yet, only the conventional foundation complies with this requirement (hence having $FS_E > 1$, as opposed to the rocking foundation which has $FS_E < 1$). Combined with the limitations on the maximum allowable eccentricity ratio (e = M/N < 2B/3), this overstrength requirement leads to the conventional foundation being significantly larger (B = 7.5 m) than the rocking-isolated one (B = 4 m). Foundation capacity was, in a first step, calculated using the well-established failure envelope relationship of [43]

$$\left(\frac{Q}{t_h}\right)^2 + \left(\frac{M}{Bt_m}\right)^2 + 2C\frac{MQ}{Bt_ht_m} = \left\{\frac{N}{N_u}(N - N_u)\right\}^2 \tag{1}$$

Where $t_h$, $t_m$, and C are parameters taken equal to 0.52, 0.35, and 0.22, respectively (determined through curve fitting of experimental results), and $N_u$ is the ultimate capacity in pure vertical loading. In addition to theoretical predictions, the foundation capacities were also calculated using numerical simulations with finite elements (FE)—details are reported in [44].

**Table 2:** Foundation design: summary of actions and factors of safety (FS)

| Breadth | B:m | 7.5 | 4 |
|---|---|---|---|
| Total vertical load | $N_{tot}$:MN | 4.9 | 4 |
| Seismic shear load | $Q_E$:MN | 0.7 | 0.7 |
| Seismic moment load | $M_E$:MNm | 7.3 | 7.3 |
| Design shear action | $Q_{Ed}$:MN | 0.7 | 0.7 |
| Design moment action | $M_{Ed}$:MNm | 7.3 | 7.3 |
| Ultimate shear capacity | $Q_u$:MN | 1.2 | 0.5 |
| Ultimate moment capacity | $M_u$:MNm | 12.9 | 4.8 |
| Factor of safety in vertical loading | $FS_V$ | 18 | 3.5 |
| Factor of safety in combined (seismic) loading | $FS_E$ | 1.77 | 0.66 |

# EXPERIMENTAL METHODS

The experimental program was carried out at the UOD and involved three different parts. First, eight RC model columns were tested in the standard four-point bending to optimize their construction procedure and verify their bending response. Second, two dynamic centrifuge model tests were carried out using elastic (aluminum) piers to measure the ultimate shear–moment capacity ($Q_u$–$M_u$) in the combined N–Q–M loading space for the two alternative foundations and check their M–θ response in comparison with FE predictions. In lieu of static pushover tests, this was achieved by exciting the soil–structure model with Ricker wavelets of suitable characteristics (PGA = 0.6 g and dominant frequency $f_E$ = 1 Hz) causing the foundations to respond well beyond their nonlinear regime. Close approximation of the monotonic M–θ backbone curve was achieved, and the results were in good agreement with FE predictions. These characterization tests are reported in [44]. The third element of testing, which is the main scope of this paper, involved four centrifuge model tests, two for each of the alternative designs, wherein the RC model piers were subjected to different earthquake scenarios using real records of varying intensities. All tests were conducted at a scale of 1:50 (n = 50) and at 50 g. The following section describes the modeling and testing procedures.

# Reinforced Concrete Column Scaled Models: Construction and Validation

Modeling cementitious material at reduced scale is prone to size effects, owing to the presence of aggregates, and can lead to significant overstrength when the scale reduces substantially (n > 10), as is usually the case in geotechnical centrifuge model testing. Therefore, this study has employed a novel scale model reinforced concrete, developed and validated by Knappett and co-workers [38, 45, 46], which realistically scales down both material stiffness and strength and reasonably replicates the response of RC structural elements under bending and shear loads. A gypsum-based mortar (beta-form surgical plaster) was used as a cementitious binder. Uniformly graded Congleton HST 95 silica sand [47] served well in modeling the aggregate phase of the concrete because its particle size distribution reasonably approximates the geometrically scaled grading curve of typical coarse aggregate at 1:50 scale. The plaster (P) and sand (S) were mixed with water (W) at a ratio of P/S/W = 1:1:1 by weight. The compressive strength of the produced concrete model was measured through tests on 100 × 100 mm standard cubes as $f_{c,}100 = 26.3$ MPa (equivalent to cylinder strength of $f'_c = 17.4$ MPa).

Reinforcement was modeled by stainless steel wire with a measured yield stress of $f_y = 460\,MPa$ (at 0.2% permanent strain), whereas the post-yield stress–strain response indicated significant hardening, making it closely representative of high yield reinforcement. The wire was cut and bent appropriately to produce scaled longitudinal bars (Figure 2(a)) and confining transverse reinforcement (Figure 2(b)). For ease of construction, wire of the same size, 0.6 mm in diameter at model scale (30-mm diameter at prototype scale), was utilized for both longitudinal and shear reinforcement. It should be noted that using shear links of smaller diameter, as is usually the case in practice, would require spacing lengths smaller than 5 mm at model scale to achieve the desired ductility, which would significantly hinder model fabrication. For both types of reinforcement, the wire was coated with HST95 silica sand using a fast-drying epoxy resin to produce a realistically rough interface between the steel and the model concrete.

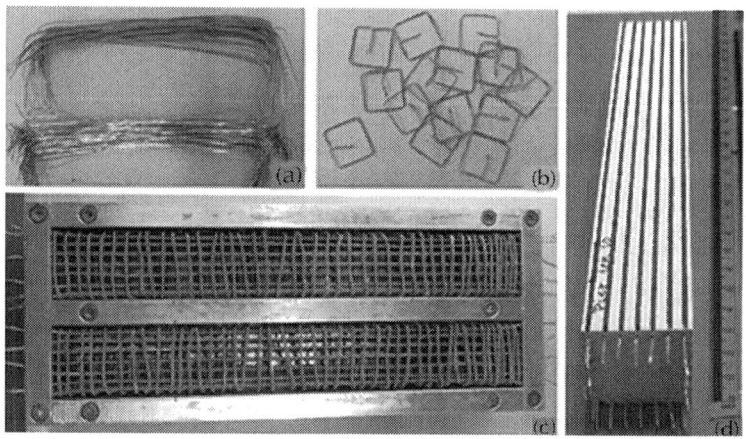

Figure 2: Construction of reinforced concrete model columns: (a) longitudinal steel with (upper) and without (lower) sand coating; (b) shear links; (c) steel within formwork; and (d) column model.

Fabrication of the reinforcement assembly was challenging because of the scale of the produced columns. The 200-mm long column model contained a total length of more than 5 m of wire modeling longitudinal reinforcement and 45 shear links uniformly spaced at a distance of approximately 5 mm. Anchoring of longitudinal reinforcement was achieved by providing an additional length of about 10 mm on each side of the column, which was bent and fixed within the foundation or the deck plates. Test units were cast in a custom-built formwork (Figure 2(c)), which allowed casting of two columns at a time. The column models (Figure 2(d)) were left to cure for at least 2 weeks before testing.

A total of 19 columns were produced, most of which were used in bending tests intended to verify the moment–curvature response of the column section and the capacity of the column–foundation joint. It is important to note that in addition to shear links, the alternative of using continuous spiral shear reinforcement was also investigated but was found less effective in producing sections with a constant core area resulting in less accurate replication of the cross-sectional capacity.

A series of standard four-point bending tests were carried out under displacement control. Figure 3(a) shows a typical test on a column model identical to those used in the centrifuge, highlighting the loading arrangement and the typical mode of failure. It can be seen that the observed crack pattern is typical of a column designed to fail in flexure (i.e., containing sufficient shear reinforcement to suppress shear failure). This was expected considering the calculated moment and shear capacities of the section given in Table 1. Vertical cracks may be identified within the 60-mm wide central span, which is subjected to pure bending (constant moment) while their inclination evidently changes in the two zones of combined moment☐shear loading between the load points and supports. Figure 3(b) compares the measured bending behavior of the column section (in prototype scale), in terms of bending moment (M) versus deflection ( ), with numerical predictions using USC-RC for pure bending conditions (without axial load) indicating very satisfactory agreement. Unfortunately, the available laboratory equipment did not allow testing with axial load to measure the effect of the pier weight on the section response. Yet, the M $\delta$ response of the concrete section under axial load equal to the pier weight (N = 3.4 MN) was calculated numerically and is also shown in Figure 3(b). As expected, the weight of the pier, which is present in the centrifuge model tests, results in considerable increase of the section moment capacity, although at the cost of reduced ductility. It is important to note that the USC-RC-predicted moment capacity for the actual axial load is in good agreement with the maximum moment load recorded in the centrifuge tests.

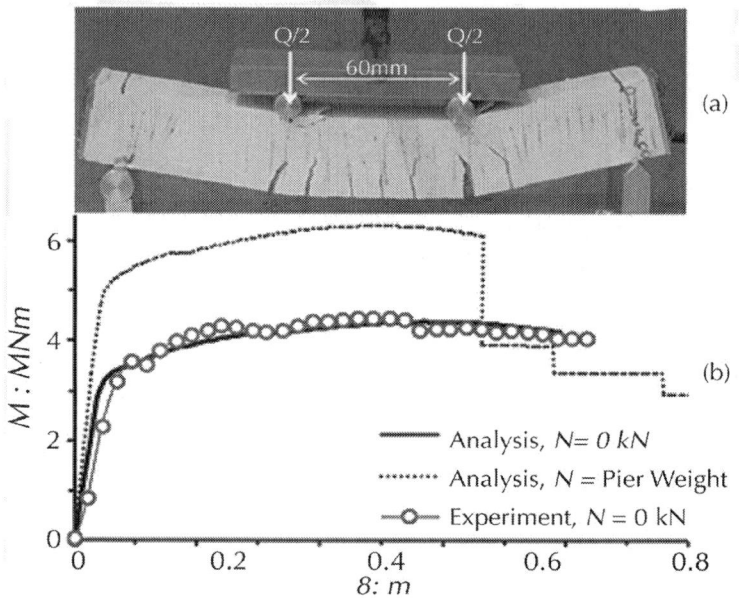

**Figure 3:** Bending response of RC column: (a) photograph of a four-point bending test on a model column; (b) comparison of computed monotonic moment–deflection response with bending test results in the absence of axial load ($N=0$ MN) and analytical prediction of over-strength due to the pier dead loads ($N=3.4$ MN).

Only the column, where structural failure was expected to take place, was modeled using the RC material, whereas the deck and the foundation were made of steel and aluminum, respectively. Figure 4(a) displays a photo of the mass-column foundation assembly indicating key dimensions. Care was taken in the design and setup of the deck-to-column and column-to-foundation joints to achieve full fixity. The elastic fixed-base vibration period of the pier models was measured through free vibration tests as $T_0 = 0.24$ s (in prototype scale), which presumably does not account for foundation geometry, being therefore the same for both design alternatives. Yet, soil-structure interaction is expected to drastically increase this value especially in the case of the rocking pier.

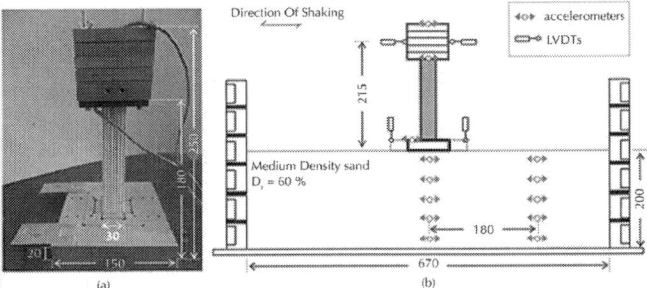

**Figure 4:** Experimental setup :(a) photo of the pier model assembly (mass, RC column, and foundation) and attached accelerometers indicating characteristic dimensions and (b) schematic cross section of the model within the laminar box showing instrumentation. All dimensions are in millimeters (model scale).

# Centrifuge Modeling

Similitude is an important consideration in physical modeling using reduced scale models that are intended to capture the response of field-scale prototypes. Centrifuge modeling is particularly useful for the investigation of soil–foundation–structure interaction problems where realistic simulation of the stress dependent soil behavior plays a key role. Thanks to the enhanced gravity field in a geotechnical centrifuge, a 1:n scale model will have the same effective stress acting at homologous points in the model soil and the full-scale prototype soil. A set of scaling laws have been developed to achieve similitude in centrifuge modeling, as detailed by [48, 49].

Dynamic centrifuge model testing was carried out in the 3.5-m diameter beam centrifuge of the UoD. A schematic cross section of the model including key dimensions and instrumentation is shown in Figure 4(b). The 200-mm deep soil layer was prepared by air pluviation of dry fine Congleton silica sand (HST95, $_{max}$=1758 kg/ m3, $y_{min}$=1459 kg/m3, $D_{60}$=0.14 mm, and $D_{10}$=0.10 mm) to achieve a uniform relative density $D_r$=60%. The soil model was prepared within an equivalent shear beam (ESB) container with flexible walls, described in [50]. It was instrumented using two vertical arrays of five ADXL78 MEMS accelerometers, one array buried under the foundation centerline and the other at a distance large enough to record free field response. The motion of the pier was recorded using identical accelerometers attached to the foundation and the deck. Vertical displacements of two foundation corners, recorded by Linear Variable Differential Transformers (LVDTs), were used to calculate settlement and rotation in the direction of shaking. Another pair of LVDTs recorded the horizontal displacements of the deck center of mass.

## Testing Protocol

Ground motions were applied using an Actidyn Q67-2 servo-hydraulic EQS. Taking advantage of the effectiveness of this device in simulating desired motions [51], an ensemble of records from historic earthquakes were utilized as base excitation. The motions were band pass filtered between 0.8 and 8 Hz to match the controllable frequency range of the EQS and then calibrated during a preliminary test on a 'dummy' model to allow highly repeatable reproduction of original records with reasonable accuracy, as illustrated by comparison of achieved and target acceleration time histories in Figure 5.

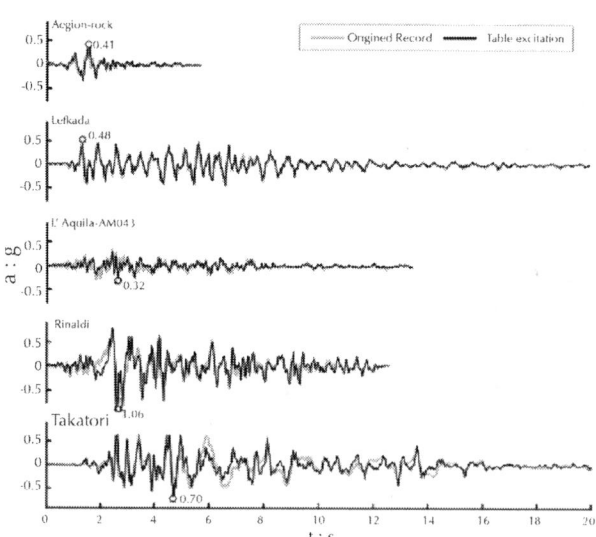

**Figure 5:** Shaking table acceleration time histories used in the different earthquake scenarios, compared with target records from earthquakes of different magnitudes in Greece, Italy, USA, and Japan.

Figure 6 shows the 5% damped acceleration response spectra of the actual shaking table motions and compares them to the EC8 design spectrum. Evidently, the Aegion-rock and L'Aquila-AM043 motions, recorded during the 1995 $M_s$ 6.2 Aegion (Greece) and the 2009 $M_s$ 6.3 L'Aquila (Italy) earthquakes, are typical medium intensity excitations of the order of magnitude assumed in design. By contrast, the Rinaldi(228) and Takatori(000) motions, recorded during the 1994 Ms6.8 Northridge (USA) and 1995 $MJ_{MA}$ 7.2 Kobe (Japan) earthquakes, are very strong motions that dramatically exceed the design level. Characterized by near fault directivity effects, these latter motions have increased spectral ordinates within a particularly broad band of periods (especially in the range of particular interest, i.e., $T \geq T_0$). Having considerable duration and number of cycles, the record of the 2003 $M_s$ 6.4 Lefkada earthquake (Greece) represents an intermediately strong, yet still above design levels, seismic scenario.

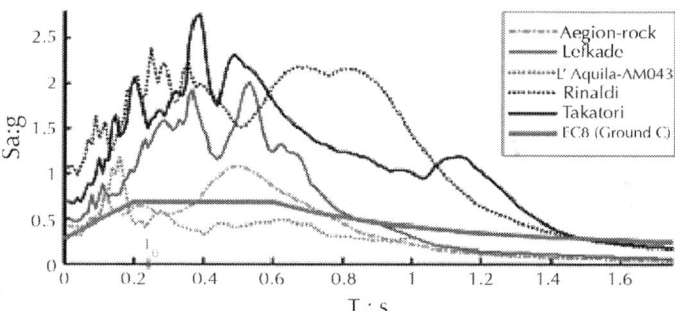

**Figure 6:** Acceleration response spectra ($\xi = 5\%$) of the shaking table excitations compared with the EC8 design spectrum for ground Type C and $a_g = 0.24$ g.

Central to the issue of seismic performance evaluation is the recognition that damage in a component is cumulative, and the level of damage depends not only on the maximum deformation but also on the history of deformations. In practice, structures may, over their lifespan, be subjected to several motions of varying intensities. Thus, attempting to account for seismic history, the performance assessment undertaken here involves a succession of motions considering two different seismic history scenarios. In Earthquake Scenario A, the ground motion order was such that the intensity roughly increased throughout the test, as if a number of weak or medium intensity motions were to precede a major catastrophic event, either in a sense of foreshocks or as independent events taking place in a relatively short period. Earthquake Scenario B explores the case where a very strong earthquake is followed by weaker shaking events or aftershocks. The two scenarios were investigated independently for each of the two designs considered with the exact testing protocol as summarized in Table 3. It should be noted that in some cases, when scenarios A or B did not lead to collapse, additional shaking was applied using the Takatori record until ultimate or practical failure was reached.

**Table 3:** Testing program: sequence of seismic excitations

| Test | Pier design | Motion protocol | | | | | | |
|------|-------------|---|---------|---------|----------|----------|----------|----------|
| 1 | Rocking isolation | A | Aegion | Lefkada | L'Aquila | Rinaldi | Takatori | — |
| 2 | Conventional design | | Aegion | Lefkada | L'Aquila | — | — | — |
| 3 | Rocking isolation | B | Rinaldi | Aegion | Aegion | L'Aquila | Takatori | Takatori |
| 4 | Conventional design | | Rinaldi | Aegion | Aegion | L'Aquila | Takatori | — |

# EXPERIMENTAL RESULTS

In this section, the seismic performance of the two pier–foundation systems during seismic scenarios A and B is evaluated and systematically compared. Focusing on damage accumulation due to successive earthquakes and the importance of shaking history, time histories of recorded illustrative demand parameters, such as deck acceleration and displacement, as well as moment-rotation and shear force-displacement hysteretic response loops are subsequently presented at prototype scale.

## Response to Earthquake Scenario A

Figure 7(a, b) shows the sequence of acceleration time histories recorded at the center of mass of the deck (the average of measurements at the top and bottom of the deck mass) in each of the two alternative designs during shaking with the first three motions of Earthquake Scenario A (plotted in Figure 7(c)). As anticipated, the rocking pier experiences invariably lower acceleration than the conventional pier. This advantage, known as rocking isolation, is the result of the difference in the ultimate moment capacity of the two designs (Table 2): the ultimate moment capacity of the RC column section is larger than the ultimate moment capacity of the rocking foundation. This advantage becomes more significant as excitation intensity increases and the ultimate capacities are mobilized. Hence, upon shaking with the L'Aquila record, the maximum transient demand experienced by the rocking pier is half the demand on the conventional pier.

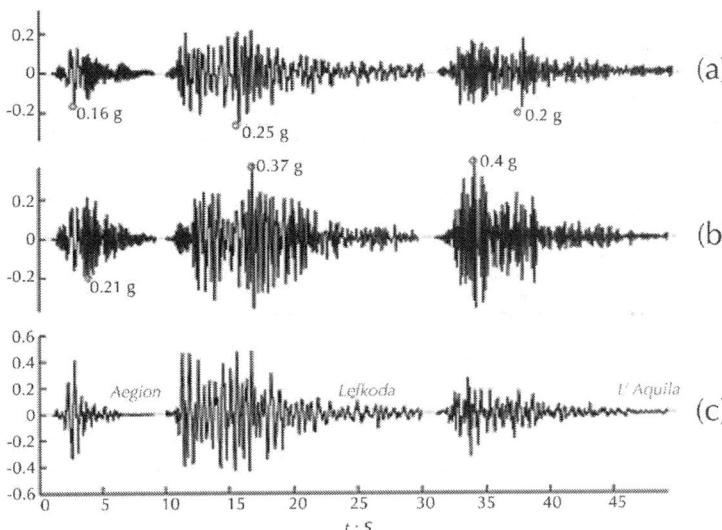

**Figure 6:** Acceleration time history sequence recorded during Earthquake Scenario A at (a) the deck of the rocking pier; (b) the deck of the conventional pier; and (c) the model base.

The rocking isolation effect can be easily quantified with regard to the moment capacity of the rocking foundation, being therefore potentially useful for the estimation of seismic demand in design. Assuming pure rotational movement about the footing midpoint, equilibrium requires the critical value of the maximum acceleration developed in the deck ($\alpha c$) to depend on the ultimate moment capacity of the foundation ($M_u$): $\alpha_c = M_u/m^{deck}h$. Given the theoretical capacity of the rocking foundation (Table 2), $a_c$ may be calculated as 0.16 g. Yet, this value coincides with the measured peak mass acceleration only in the very first shaking event using the Aegion record. Thereupon, overstrength effects, associated with soil densification during shaking, lead to some considerable increase in this value, which yet remains substantially lower than the peak demand on the conventional pier. It should be noted that deviations of the maximum deck acceleration from the theoretical αc value may be also because of uncertainties in the estimation of $M_u$ and the simplifying presumption of pure rotational movement. Nevertheless, such overstrength effects, which lead to some considerable increase of $M_u$ and $\alpha_c$ due to preceding loading cycles, have been identified in the past experimental studies and documented in [23, 30].

On the basis of the aforementioned discussion, it may be deduced that even the relatively low intensity Aegion excitation is sufficient to momentarily mobilize the capacity of the rocking footing and induce its rocking–uplifting response. Indeed, as indicated by the deck displacement time history plots in Figure 8(a), the pier response is primarily controlled by rotational movement due to foundation rocking. Horizontal deck displacement measurements (Figure 8) are separated into their two main components: the lateral displacement due to foundation rotation ($\delta_\theta$) and due to the flexural deformation of the pier ($\delta_{col}$). The evolution of $\Delta$ throughout shaking serves as an index of the two systems seismic performance. Moreover, superimposing the contribution of $\delta_\theta$ and $\delta_{col}$ is intended to illustrate the dominating mode of response. It should be noted that, unlike $\Delta$ and $\delta_\theta$, $\delta_{col}$ was not measured directly but calculated with respect to the measurements of the other two as $\delta_{col} = \Delta - \delta_\theta$, assuming that for such slender oscillators, foundation sliding ($\delta_s$) is minimal. Evidently, as anticipated, in the case of the rocking pier, rotational movement prevails throughout the entire shaking sequence. The opposite is the case with the conventional pier, where deck drift is almost exclusively associated with deformation and failure of the RC column. More importantly, the rocking pier demonstrates a crucial advantage over conventional design. Not only does it experience significantly lower drift in the first shaking event (Aegion) but it also retains an increasingly favorable performance during the following earthquakes of greater intensity and duration. In particular, having experienced the first two excitations, it is practically unaffected by the sustained shaking of the L'Aquila record, resulting in a small total drift of 0.1 m (≈0.1% drift ratio). In contrast, the conventional pier accumulates large flexural deformations leading to approximately three times the deck drift seen for the rocking pier after the second motion and eventually failing with the third motion, having acquired drift levels of more than 50 cm (or≈0.5% drift ratio). Note that in this test, failure is assumed to take place when the deck mass hits the horizontal LVDT, which prevents further movement in the direction of shaking. Yet, the rate at which $\Delta$ is observed to increase just before hitting the instrument (Figure 8(b)) implies imminent collapse of the pier column.

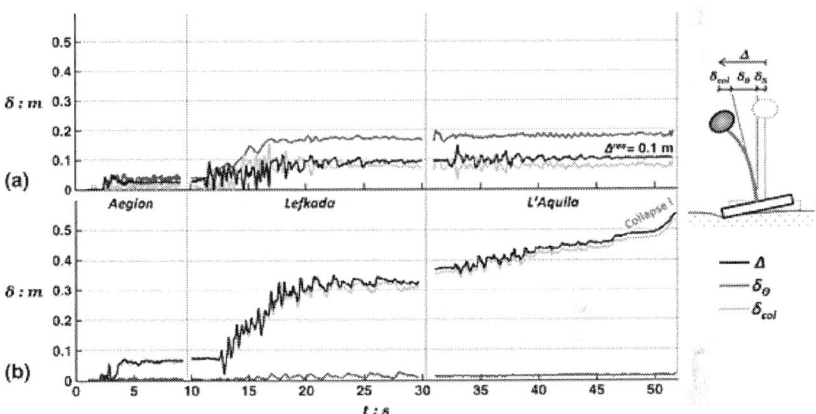

**Figure 8:** Total deck drift $\Delta$ of (a) the rocking pier and (b) the conventional pier, shown as the components of rotational movement $\delta_\theta$ and flexural deformation $\delta_{col}$ during shaking with Earthquake Scenario A.

Inelastic action is not only unavoidable but also essential in providing the required energy absorption to enable survival of the structure under intense seismic shaking. The two design alternatives, both relying on inelastic response, differ only in the component where the inelastic deformation is directed to. The adequacy of these nonlinear components ('fuses') may refer to their strength, ductility, cumulative damage resistance, and energy dissipation capacity. The effectiveness of the designated 'fuses', either the RC column section for conventional design or the soil–foundation interface for the rocking design, is of critical importance for the survivability of the system. Direct comparison of the two piers' 'fuse' response is highlighted by comparing the family of hysteresis plots depicted in Figure 9.

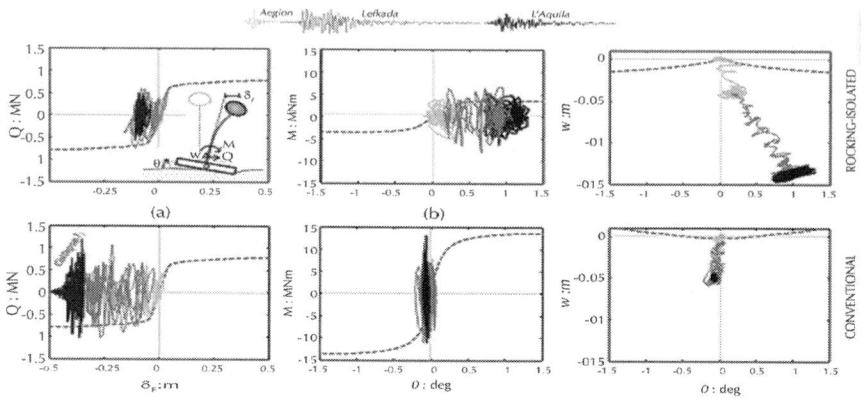

**Figure 9:** Rocking (top) versus conventional (bottom) pier hysteretic responses: (a) Q– $_{col}$ with reference to the base of the RC column; (b) foundation M–θ;
and (c) foundation w– for Earthquake Scenario A.

Figure 9(a) shows the response of the RC column during the three
first earthquakes of Scenario A in terms of shear load (Q = m ) versus the
flexural component of drift ( col). It can be seen that the conventional
pier column marginally mobilizes its lateral load capacity during the
Aegion earthquake resulting in some rather minimal permanent flexural
deflection ($\delta_{co}$l/h < 0.1%). Subsequent loading with the stronger Lefkada
motion causes an important number of excursions into the nonlinear
regime, accumulating substantial permanent deformation at the column
base. Nevertheless, designed according to modern code requirements,
the pier column possesses adequate confinement and, hence, ductility
to sustain such a significant number of loading cycles with no apparent
deterioration of strength, yet at the cost of considerable permanent
deflection $\delta_{col} \approx 30$ cm ($\delta_{col}$/h ≈ 0.3%). However, having experienced this
damage, the pier column appears unable to sustain further excitation
with the equally strong L'Aquila motion, which exhausts its ductility
capacity causing rapid deterioration of strength after a couple of cycles
and eventually collapse. On the other hand, the column of the rocking
pier responds, as expected, practically within the linear-elastic regime
throughout the entire sequence.

Figure 9(b,c) summarizes the performance of the two foundations
in the moment-rotation and settlement-rotation domains. Verifying
its design, the conventional foundation responds linear elastically
with increased rotational stiffness, in comparison with the monotonic

backbone curve, owing to densification of the underlying soil. By contrast, the rocking foundation presents a broad moment–rotation hysteresis receiving comparatively larger rotational demand. Nevertheless, rotational movement is kept within tolerable margins, as may be judged with respect to the resulting deck deflections (Figure 8). Downwards movement prevails resulting in considerable settlement of the foundation from the first earthquake ($\approx$5 cm, which is equal to the total settlement of the conventional foundation over the complete scenario), this being the main shortcoming of the rocking design. Foundation settlement increases drastically in response to the multiple shaking cycles of the Lefkada record, leading to a considerable amount of settlement (w$\approx$14 cm) by the end of the third earthquake. Nevertheless, despite the increased settlements, the rocking pier appears to have a crucial advantage over the conventional pier: not only does it avoid collapse but it also exhibits a particularly effective and ductile 'fuse' mode of response, as its rocking foundation sustains a large number of loading cycles with no apparent deterioration of strength.

Although the collapse of the conventional pier column after excitation with the L'Aquila record ended Test 2, Test 1 was continued by applying two additional very strong motions, the Rinaldi and the Takatori records. Figure 10 shows the time history response of the rocking pier during these last, particularly intense, excitations in terms of deck drift, foundation settlement, and deck acceleration.

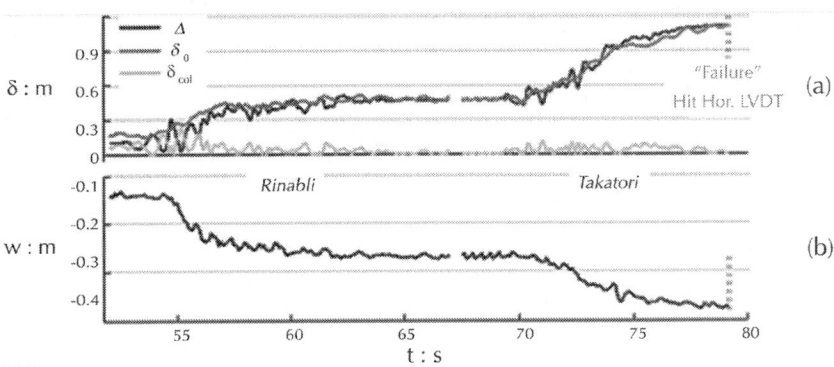

**Figure 10:** Response of the rocking pier subjected to successive base excitation with the Rinaldi and the Takatori motions, after having survived shaking

with the three preceding lower magnitude motions of Earthquake Scenario A, in terms of (a) deck drift and (b) foundation settlement time histories.

Remarkably, despite having been subjected to a sequence of three earthquakes with intensity equivalent to, or exceeding, its design earthquake and having suffered considerable foundation deformation, the rocking pier survives the excess demands imposed by the Rinaldi motion. Displacements, in both horizontal and vertical directions, are naturally increased substantially (wres = 27 cm, $\delta_{tot}$ = 45 cm), but the response is judged as satisfactory as the pier remains stable after such a deleterious sequence of earthquakes. The pier eventually failed (again by hitting the horizontal LVDT) during excitation with the last, extremely strong excitation using the Takatori record. Figure 11 shows images of the two models after testing, verifying the respective mode of response.

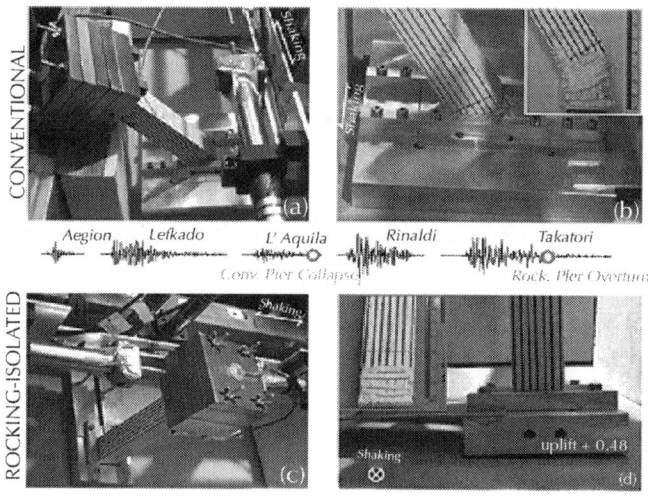

**Figure 11:** Photos of the bridge models after Tests 1 and 2 (Scenario A): (a) failure of the conventional pier after the first three medium-strong intensity motions and (b) the damage at the pier base compared with (c) the rocking pier subjected to the same motions plus two additional very strong motions, focusing on (d) foundation uplift.

The realistic reproduction of flexural failure in the case of the conventional pier (Figure 11(b)) is worth observing, where the plastic hinge length is approximately equal to the column width, as expected in practice. Note, also, that after hitting the horizontal LVDT during the L'Aquila motion, the pier was found to have collapsed in the out-of-plane direction (Figure 11(a,b)). After being subjected to the entire sequence of Scenario A (five earthquakes), the rocking pier rotated significantly (Figure 11(c)) with evident foundation uplift, yet with its RC column remaining practically intact (Figure 11(d)).

# Response to Earthquake Scenario B

The response of nonlinear systems strongly depends on the exact loading history. Hence, it was decided to further study the response of the two pier designs under an alternative earthquake sequence in order to generalize the previously made observations. This loading scenario differs from Earthquake Scenario A in that the very intense Rinaldi record is applied first, whereas the weaker Aegion and Lefkada records, subsequently imposed on the models, may be perceived as smaller aftershocks. Figure 12 shows the acceleration time history of the first four motions involved in this scenario (note that the Aegion record was applied twice) in comparison with the accelerations recorded at the decks of the two piers. Measurements are in qualitative agreement with those in Figure 7 verifying the beneficial effect of rocking isolation in drastically reducing seismic demands on the rocking pier.

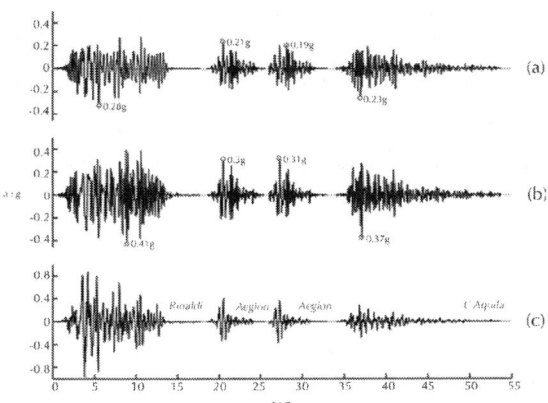

**Figure 12:** Acceleration time history sequence recorded during shaking with Earthquake Scenario B at (a) deck of rocking pier (smaller foundation); (b) deck of conventional pier (larger foundation); and (c) excitation.

Figure 13 shows the evolution of deck drift for each of the pier models highlighting the contribution of rotational movement ($\delta\Theta$) and column deflection ($\delta$col). Again, in agreement with results from Tests 1 and 2, it may be seen that for the rocking pier, deck deflections are mainly because of the foundation rotation, whereas column deflection plays a minor role. The opposite is the case for the conventional pier. The rocking pier suffers significantly less total drift than the conventional pier in the Rinaldi earthquake (31 cm, or $\Delta/h \approx 0.3\%$, against 47 cm, or $\Delta/h \approx 0.4\%$). Despite being considerably distressed by the first very strong motion, the rocking pier demonstrates a remarkably stable response during shaking with the following three smaller excitations. Showing surprising resistance against cumulative damage, the rocking foundation suffers little additional rotation and hence negligible additional deck displacement (Figure 13(a)). In contrast, each shaking event adds considerably to the deflection of the conventional pier (Figure 13(b)). By the end of Earthquake Scenario B, the conventional pier was subjected to twice the deck drift ($\approx$63 cm) associated with the rocking alternative.

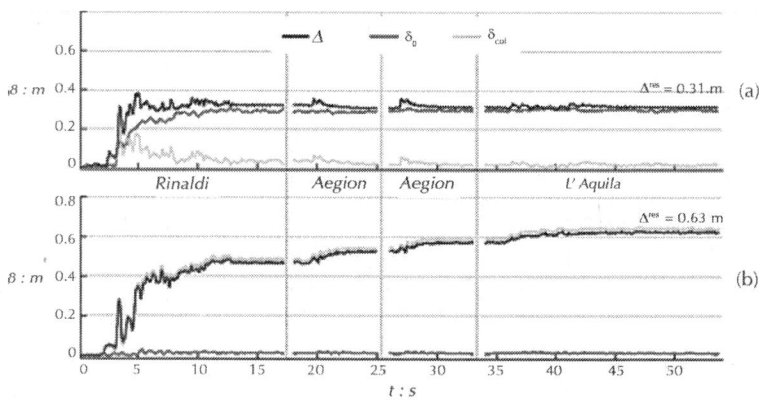

**Figure 13:** Total deck drift $\Delta$ of (a) rocking pier and (b) the conventional pier, shown as the components of rotational movement $\delta\Theta$ and flexural deformation col during shaking with Earthquake Scenario B.

Comparison of hysteretic responses recorded during the Rinaldi earthquake (Figure 14) shows strongly nonlinear behavior of the RC column or the foundation in the case of the conventional design or the rocking design, respectively. It is worth observing that this first strong motion pushes the response of the conventional column (Figure 14(a)) well within its nonlinear regime consuming more than half of its theoretical ductility capacity. The following lower magnitude earthquakes add up to the total ductility demand and the column is observed to be 'on the verge of failure' by the end of the shaking sequence, having exhausted its theoretical ductility margins. On the other hand, the rocking foundation exhibits good energy dissipating behavior with no deterioration of foundation capacity (Figure 14(b)). Furthermore, it is important to observe that, as with horizontal displacements, there is negligible additional settlement of the rocking foundation due to the post-Rinaldi earthquake loading (Figure 14(c)).

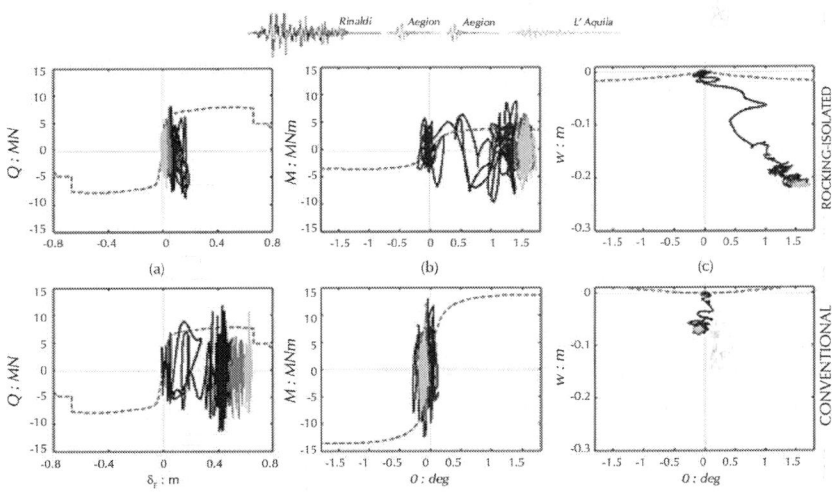

**Figure 14:** Rocking (top) versus conventional (bottom) pier hysteretic responses: (a) Q–δcol with reference to the base of the RC column; (b) foundation M–θ; and (c) foundation w–θ for Earthquake Scenario B.

Because both piers survived Earthquake Scenario B without collapse, testing was continued by applying the deleterious Takatori motion. Figure 15 shows the recorded displacements at the deck of each pier. Demonstrating surprising resistance, the rocking pier survives this extreme shaking reaching a permanent deck drift due to rotation (rather than pier flexure) of 58 cm (note that this value is lower than the total drift of the conventional pier after the preceding shaking events). In fact, the second shaking with the Takatori excitation was required to induce failure of the rocking pier. In contrast, the conventional pier failed just after the application of the first Takatori motion. Figure 16 shows images of the conventional pier model after testing, showing the significant structural damage at the base of its RC column, in this case, in the plane of shaking. Again, the damage pattern (hinge length, crack formation, and compressive spalling) implies realistic modeling of the actual response of RC elements.

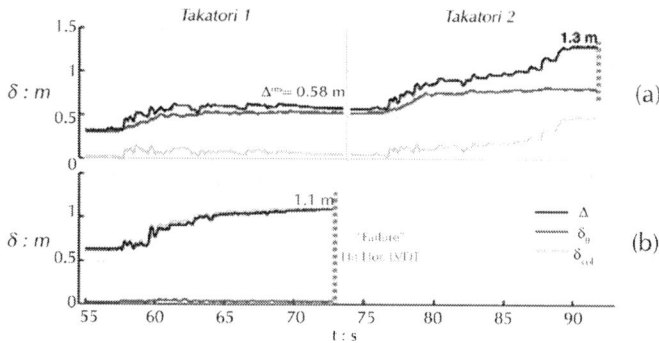

**Figure 15:** Drift response of (a) the rocking pier and (b) the conventional pier to base excitation with two successive Takatori records following the shaking sequence of Earthquake Scenario B.

**Figure 16:** Photos of the conventional pier after Test 6 indicating the damage induced by shaking with Earthquake Scenario B followed by one additional very strong motion (the Takatori record): (a) view of the entire pier model (b) in-plane and (c) out-of-plane views of the column–foundation joint.

# CONCLUDING REMARKS

This experimental campaign has provided proof of the concept of deliberately designing for foundation nonlinearity to render RC structures safe under intense seismic excitation. Emphasis is placed on the response of RC bridge piers designed in accordance with modern seismic codes and hence having well-confined cross sections. In contrast to previous studies, these piers are modeled at a highly reduced scale (1:50) using a recently developed scale model reinforced concrete, which captures the behavior of prototype RC sections with a high level of fidelity. The seismic performance of a moderately tall pier supported on a square footing on top of a layer of medium-dense sand was studied through a series of dynamic centrifuge experiments. Two design alternatives are considered: (i) conventional capacity design, in which the foundation is as usual overdesigned, guiding plastic hinging into the superstructure and (ii) rocking isolation design, in which the foundation is deliberately underdesigned to promote uplifting and soil yielding, guiding plastic deformation below the ground. The performance of the two design alternatives is evaluated and compared with the emphasis on the resistance of potential plastic hinge zones to cumulative damage due to successive shaking protocols. The testing sequence involved two different shaking scenarios, representing particularly destructive earthquake motions either preceded or followed by earthquakes closer to those against which the pier was designed.

On the basis of the presented tests, the following conclusions can be drawn:

1.  The utilized small-scale concrete model simulates the behavior of a prototype RC section reliably, allowing greatly improved prediction of the detailed nonlinear seismic response of RC structures.

2.  Rocking isolation design consistently exhibits superior performance compared with conventional capacity design, irrespective of the tested shaking scenario. Nonlinear response of the soil-footing interface essentially acts as an effective and resilient energy dissipation 'fuse' showing substantially increased ductility capacity and resistance against cumulative damage despite the successive and intense seismic shaking.

In both shaking scenarios, the rocking-isolated pier survived the excessive demands imposed by particularly strong shaking sequences, which caused catastrophic structural failure of the conventionally designed pier.

3. Counterintuitively, the rocking-isolated pier was found to be advantageous also in terms of drift demands suffering comparatively lower deck displacements in all of the studied loading cases.

4. Exhibiting what is known as a 'sinking response', the rocking foundation was found to accumulate significant settlement, this being identified as the only drawback of the rocking isolation design. Naturally, owing to its significantly lower FSV, the rocking foundation is prone to suffering increased settlements in comparison with the overdesigned foundations involved in conventional capacity design. The effect of such settlements on performance of the bridge could not be measured in the presented tests, where the continuity of the deck and the interaction between consecutive piers have not been taken into account. Yet, it should be noted that settlements are expected to be significantly limited should the pier be founded on a denser soil stratum ($Dr > 60\%$), dense enough to actually support the option of a shallow foundation or be subjected to a less deleterious shaking. Moreover, this drawback may be remediated through soil improvement measures. In fact, previous experimental studies on rigid oscillators [52] have shown that improvement, that is, densification of a soil layer of depth equal to the foundation width (4 m deep in this particular case) may drastically reduce rocking-induced foundation settlements, thereby alleviating concerns about the idea of implementing rocking isolation in practice.

# ACKNOWLEDGMENTS

This work was financially supported by the European Research Council (ERC) Programme IDEAS, Support for Frontier Research—Advanced Grant, under Contract number ERC–2008–AdG 228254–DARE. The authors gratefully acknowledge the support of the manager of the UoD centrifuge facility, Dr Andrew Brennan, and the invaluable contribution of the technical staff within the Department of Civil Engineering, especially Mark Trusswell and Colin Stark, who assisted in conducting the experiments. Moreover, thanks are due to Asad Al-Defae, Patrick Madden, and Daniele Bertalot for their assistance with the centrifuge testing work.

# REFERENCES

1.  Paulay T, Priestley MJN. Seismic design of reinforced concrete and masonry buildings. Wiley: New York, 1992.

2.  Villaverde R. Methods to assess the seismic collapse capacity of building structures: state of the art. Journal of Structural Engineering, ASCE 2007; 133(1):57–66.

3.  Smyrou E, Tasiopoulou P, Bal IE, Gazetas G. Ground motions versus geotechnical and structural damage in the Christchurch February 2011 earthquake. Seismological Research Letters 2011; 82(6):882–892

4.  Housner GW. The behavior of inverted pendulum structures during earthquakes. Bulletin of the Seismological Society of America 1963; 53(2):404–417.

5.  Meek JW. Effect of foundation tipping on dynamic response. Journal of the Structural Division, ASCE 1975; 101(7):1297–1311.

6.  Yim SC, Chopra AK. Simplified earthquake analysis of structures with foundation uplift. Journal of Structural Engineering, ASCE 1985; 111(4):906–930.

7.  Paolucci R. Simplified evaluation of earthquake induced permanent displacements of shallow foundations. Journal of Earthquake Engineering 1997; 1(3):563–579.

8.  Yim CS, Chopra AK, Penzien J. Rocking response of rigid blocks to earthquakes. Earthquake Engineering and Structural Dynamics 1980; 8:565–587.

9.  Shenton HW. Criteria for initiation of slide, rock, and slide-rock rigid-body modes. Journal of Engineering Mechanics, ASCE 1996; 122(7):690–693.

10. Makris N, Roussos YS. Rocking response of rigid blocks under near-source ground motions. Geotechnique 2000; 50(3):243–262.

11. Zhang J, Makris N. Rocking response of free-standing blocks under cycloidal pulses. Journal of Engineering Mechanics, ASCE 2001; 127(5):473–483.

12. Makris N, Konstantinidis D. The rocking spectrum and the limitations of practical design methodologies. Earthquake Engineering and Structural Dynamics 2003; 32, 265–289.

13. Gerolymos N, Apostolou M, Gazetas G. Neural network analysis of the overturning response under near–fault type excitation. Earthquake Engineering and Engineering Vibration 2005; 4(2):213–228.

14. Apostolou M, Gazetas G, Grini E. Seismic response of slender rigid structures with foundation uplift. Soil Dynamics and Earthquake Engineering 2007; 27:642–654.

15. Mergos PE, Kawashima K. Rocking isolation of a typical bridge pier on spread foundation. Journal of Earthquake Engineering 2005; 9(2):395–414.

16. Pecker A. Design and construction of the foundations of the RionAntirion Bridge. Proceedings of the 1st Greece–Japan Workshop on Seismic Design, Observation, Retrofit of Foundations, Athens, 2005; 119–130.

17. Chen YH, Liao WH, Lee CL, Wang YP. Seismic isolation of viaduct piers by means of a rocking mechanism. Earthquake Engineering and Structural Dynamics 2006; 35:713–736.

18. Faccioli E, Paolucci R, Vanini M. TRISEE: 3D site effects and soil-foundation interaction in earthquake and vibration risk evaluation. European Commission Publications, 1999.

19. Martin GR, Lam IP. Earthquake resistant design of foundations: retrofit of existing foundations. Proceedings of GeoEng 2000 Conference, Melbourne, 2000; 19–24.

20. Gajan S, Kutter B, Phalen J, Hutchinson T, Martin G. Centrifuge modeling of load-deformation behavior of rocking shallow foundations. Soil Dynamics and Earthquake Engineering 2005; 25:773–783.

21. Shirato M, Kouno T, Asai R, Nakani N, Fukui J, Paolucci R. Large-scale experiments on nonlinear behavior of shallow foundations subjected to strong earthquakes. Soils and Foundations 2008; 48(5):673–692.

22. Gajan S, Kutter B. Capacity, settlement, and energy dissipation of shallow footings subjected to rocking. Journal of Geotechnical and Geoenvironmental Engineering, ASCE 2008; 135(3):407–420.

23. Drosos V, Georgarakos P, Loli M, Zarzouras O, Anastasopoulos I, Gazetas G. Soil–foundation–structure interaction with mobilization of bearing capacity: an experimental study. Journal of Geotechnical and Geoenvironmental Engineering, ASCE 2012; 138(11):1369–1386.

24. Chang BJ, Raychowdhury P, Hutchinson TC, Thomas J, Gajan S, Kutter B. Evaluation of the seismic performance of combined frame-wall-foundation structural systems through centrifuge testing. Proceedings of the 4th International Conference on Earthquake Geotechnical Engineering (4ICEGE), Thessaloniki, Greece, 2007.

25. Gelagoti F, Kourkoulis R, Anastasopoulos I, Gazetas G. Rocking isolation of low-rise frame structures founded on isolated footings. Earthquake Engineering and Structural Dynamics 2012; 41(7):1177–1197.

26. Anastasopoulos I, Gelagoti F, Spyridaki A, Sideri J, Gazetas G. Seismic rocking isolation of asymmetric frame on spread footings. Journal of Geotechnical and Geoenvironmental Engineering 2014; 140(1):133–151.

27. Anastasopoulos I, Gazetas G, Loli M, Apostolou M, Gerolymos N. Soil failure can be used for seismic protection of structures. Bulletin of Earthquake Engineering 2010; 8(2):309–326.

28. Hung H, Liu K, Ho T, Chang K. An experimental study on the rocking response of bridge piers with spread footing foundations. Earthquake Engineering and Structural Dynamics 2011; 40(7):749–769.

29. Deng L, Kutter B, Kunnath S. Centrifuge modeling of bridge systems designed for rocking foundations. Journal of Geotechnical and Geoenvironmental Engineering, ASCE 2012; 138(3):335–344.

30. Anastasopoulos I, Loli M, Georgarakos T, Drosos V. Shaking table testing of rocking–isolated bridge pier on sand. Journal of Earthquake Engineering 2013; 17(1):1–32.

31. Allotey N, El Naggar MH. An investigation into the Winkler modelling of the cyclic response of rigid footings. Soil Dynamics and Earthquake Engineering 2008; 28(1):44–57.

32. CrossRef

33. Paolucci R, Shirato M, Yilmaz MT. Seismic behaviour of shallow foundations: shaking table experiments vs numerical modeling. Earthquake Engineering and Structural Dynamics 2008; 37:577–595.

34. Chatzigogos CT, Pecker A, Salencon J. Macroelement modelling of shallow foundations. Soil Dynamics and Earthquake Engineering 2009; 29:765–781.

35. Gajan S, Kutter BL. Effects of moment-to-shear ratio on combined cyclic load-displacement behavior of shallow foundations from centrifuge experiments. Journal of Geotechnical and Geoenvironmental Engineering, ASCE 2009; 135(8):1044–1055.

36. Raychowdhury P, Hutchinson TC. Performance evaluation of a nonlinear Winkler-based shallow foundation model using centrifuge tests results. Earthquake Engineering and Structural Dynamics 2009; 38(5):679–698.

37. Anastasopoulos I, Gelagoti F, Kourkoulis R, Gazetas G. Simplified constitutive model for simulation of cyclic response of shallow foundations: validation against laboratory tests. Journal of Geotechnical and Geoenvironmental Engineering, ASCE 2011; 137(12):1168–1154.

38. Figini R, Paolucci R, Chatzigogos CT. A macro-element model for non-linear soil–shallow foundation–structure interaction under seismic loads: theoretical development and experimental validation on large scale tests. Earthquake Engineering and Structural Dynamics 2011; 44(3):475–493.

39. Knappett JA, Reid C, Kinmond S, O'Reilly K. Small-scale modelling of reinforced concrete structural elements for use in a geotechnical centrifuge. Journal of Structural Engineering 2011; 137(11):1263–1271.

40. Trombetta NW, Mason HB, Chen Z, Hutchinson TC, Bray JD, Kutter BL. Nonlinear dynamic foundation and frame structure response observed in geotechnical centrifuge experiments. Soil Dynamics and Earthquake Engineering 2013; 50:117–133.

41.   CEN. Eurocode 2: design of concrete structures — part 2: concrete bridges. Design and detailing rules. EN 1992-2:2005. Brussels, 2005.

42.   CEN. Eurocode 8: design of structures for earthquake resistance — part 1: general rules, seismic actions and rules for buildings. EN 1998-1:2004. Brussels, 2004.

43.   Esmaeily GA, Xiao Y. Seismic behavior of bridge columns subjected to various loading patterns. University of California, Berkeley, PEER Report 2002/15, 2002.

44.   Butterfield R, Gottardi G. A complete three dimensional failure envelope for shallow footings on sand. Geotechnique 1994; 44(1):181–184.

45.   Loli M, Anastasopoulos I, Knappett J, Brown M. Use of Ricker wavelet ground motions as an alternative to push-over testing. Proc. 8th International Conference on Physical Modelling in Geotechnics, ICPMG'14, Perth, Australia, January 14–17, 2014; 1073–1078.

46.   Al-Dafae AH, Knappett JA. Centrifuge modelling of the seismic performance of pile-reinforced slopes. Journal of Geotechnical and Geoenvironmental Engineering 2014; 140(6), Paper number 04014014. DOI: 10.1061/(ASCE)GT.1943-5606.0001105)

47.   Al-Defae AH, Knappett JA. Stiffness matching of model reinforced concrete for centrifuge modelling of soil-structure interaction. Proc. 8th International Conference on Physical Modelling in Geotechnics, ICPMG'14, Perth, Australia, January 14–17, 2014; 1067–1072.

48.   Al-Defae AH, Caucis K, Knappett JA. Aftershocks and the whole life seismic performance of granular slopes. Geotechnique 2013; 63(14):1230–1244.

49.   Schofield AN. Cambridge geotechnical centrifuge operations. Geotechnique 1980; 30(3):227–268.

50.   Kutter BL. Recent advances in centrifuge modeling of seismic shaking. In: Proceedings of 3rd International Conference on Recent Advances in Geotechnical Earthquake Engineering and Soil Dynamics, April 2– 7, St. Louis, Missouri, 1995.

51. Bertalot D. Behaviour of shallow foundations on layered soil deposits containing loose saturated sands during earthquakes. PhD thesis, University of Dundee, UK, 2013.

52. Brennan AJ, Knappett JA, Bertalot D, Loli M, Anastasopoulos I, Brown M. Dynamic centrifuge modelling facilities at the University of Dundee and their application to studying seismic case histories. Proc. 8th International Conference on Physical Modelling in Geotechnics, ICPMG'14, Perth, Australia, January 14–17, 2014; 227–233.

53. Anastasopoulos I, Kourkoulis R, Gelagoti F, Papadopoulos E. Response of SDOF systems on shallow improved sand: an experimental study. Soil Dynamics and Earthquake Engineering 2012; 40:15–33.

# Centrifuge Modelling of a Soil Nail Retaining Wall

Prof SW Jacobsz

Department of Civil Engineering, University of Pretoria Pretoria, 0002.

# ABSTRACT

This paper describes a physical model of a soil nail retained excavation face which was tested in the new geotechnical centrifuge at the University of Pretoria. As centrifuge modelling is new in South Africa, a short introduction to this technique is presented. The mobilisation of soil nail forces and their maximum values in response to excavation in the model were compared to measurements recently made in an instrumented 10 m high soil nail retaining structure for the Gautrain system in Pretoria. Results were also compared to predictions made using a simple failure wedge analysis and a database of eleven full-scale instrumented soil nail walls from the literature. The centrifuge model data compared well with both full-scale situations and theoretical analyses. The results suggest that soil nail forces measured in the centrifuge are conservative due to the mobilisation of a portion of the shear strength of the model soil during the acceleration of the centrifuge, leaving less un-mobilised shear strength available to resist loads resulting from the excavation.

# INTRODUCTION

Various analytical methods can be used to assess collapse loads of geotechnical problems, e.g. plasticity solutions like the slip-line method or the limit equilibrium methods which have traditionally been the most widely used method (Shen *et al* 1982). However, limit equilibrium methods require assumptions regarding the shape of the failure surface and the distribution of stress along the failure surface. As these assumptions affect the solution of the problem, it is important that they are realistic. Failure mechanisms and deformation behaviour of soil-nailed structures can be back-analysed from full-scale case studies, which are rare and costly, or from laboratory model studies. The non-linear stress-strain properties of soils require the stress levels in models to be corrected to that of the full scale to ensure realistic results. This necessitates the use of a geotechnical centrifuge.

Shen *et al* (1982) reported on one of the first centrifuge model studies conducted to model a soil nail retaining wall in sand and compared test results against the predictions from analytical models. A comprehensive study of soil-nailed walls in sand was also carried out by Tei (1993). Zhang *et al* (2001) carried out parametric studies of soil nail retaining structures, experimenting with nail lengths and spacings, and found that failure surfaces of nailed surfaces were deeper than without reinforcement. Shen *et al* (1982) and Tei (1993) observed curved failure wedges (logarithmic spirals, according to Tei *et al* 1998; see also Bolton & Pang 1982), initiating from the toe of the retained face and reported good agreement with critical failure wedges predicted from limit equilibrium analysis.

Physically modelling all elements of the process of constructing a soil nail retained face in the centrifuge presents many difficulties. In the available case studies, the soil nails were pre-installed during model preparation. Modelling of the excavation can, however, be achieved relatively easily by draining a fluid selected to exert a horizontal pressure approximately equal to that of the soil once the desired acceleration had been achieved (e.g. Tei 1993). Other researchers did not model the excavation process and simply accelerated the completed model to the required acceleration (e.g. Shen *et al* 1982 and Zhang *et al* 2001). Despite some obvious discrepancies, both reported the performance of the model to be comparable to that of the full-scale situation yielding realistic results.

# The Geotechnical Centrifuge

The Department of Civil Engineering at the University of Pretoria, South Africa, has recently acquired a geotechnical centrifuge with a capacity of 150 G-ton, meaning that the centrifuge is capable of accelerating a payload weighing up to one ton to 150 G. Geotechnical centrifuges are used to subject small-scale models of geotechnical situations to high accelerations. Due to the stress-strain behaviour of soils being highly non-linear, it is necessary to increase the stresses in a model to be analogous to the stress distribution in the full-scale situation. This is achieved using centripetal acceleration. As such, a model with a scale of 1:50 has to be accelerated to 50 times earth's gravity (50 G) to create the correct stress distribution.

Model dimensions scale linearly and can be used to derive scaling laws for other physical properties. Table 1 lists scaling laws for a number of physical quantities. As an example, the scaling law for force is derived:

**Table 1**: Scaling laws for various physical properties

| Model scale | n |
|---|---|
| Accelerations | n |
| Linear dimensions | 1/n |
| Stress | 1 |

| Strain | 1 |
|---|---|
| Density | 1 |
| Mass | $1/n^3$ |
| Force | $1/n^2$ |
| Bending moment | $1/n^3$ |
| Moment of area | $1/n^4$ |
| Time (consolidation) | $1/n^2$ |
| Time (dynamic | $1/n$ |
| Time (creep) | $1/n$ |
| Pore fluid velocity | $n$ |

According to Newton's second law, force $(F_p)$ in the full-scale situation (the prototype) can be expressed as $F_p = m_p a_p$, where $m_p$ is the mass and $a_p$ the acceleration of the prototype. Assuming that the body to be scaled is a cube with density $\rho$ and side length $l_p$ and that it is stationary on the earth's surface, Newton's second law can be written as

$$F_p = \rho l_p^3 g \tag{1}$$

where $g$ is gravitational acceleration. Newton's second law for the model is

$$F_p = \rho l_p^3 g \tag{2}$$

where $F_m$ is force at the model scale, $m_m$ the mass of the model and $a_m$ the acceleration at model scale. In order to avoid problems with different material properties, the same material as that occurring in the full-scale situation is normally used to create the model. The material density ( ) therefore remains the same. The model is $N$ times smaller than the prototype and is therefore accelerated to $N$ times earth's gravitational acceleration to create the correct stress distribution in the model. Equation 2 therefore becomes

$$F_p = \rho l_p^3 g \tag{3}$$

which proves the scaling law for force.

$$F_m = \rho \left(\frac{l_p}{N}\right)^3 Ng = \frac{\rho l_p^3 g}{N^2} = \frac{F_p}{N^2}$$

In terms of scaling laws, particularly attractive is the fact that time-related problems, e.g. consolidation, may be studied in a fraction of the time that would be required for a full-scale trial. Also, stiffnesses (e.g. the Young's and shear moduli) do not scale because stresses and strains do not scale. This enables the same material from the full-scale prototype to be used to construct the model.

Jacobsz & Phalanndwa (2011) described a case study in which three instrumented soil nails were installed in a retained face along a cutting for the Gautrain railway line in Pretoria. The structure was excavated in residual andesite which increased in strength and stiffness with depth. The wall was 10 m high with six rows of nails installed at vertical spacings of 1.5 m and horizontal spacings of 2 m, and at a downward angle of 10°. The shotcrete facing was 175 mm thick, reinforced with two layers of mesh. The retained face and the locations of the instrumented couplings are illustrated in Figure 1. Axial forces in three of the soil nails were measured as the excavation in front of the retained face was deepened.

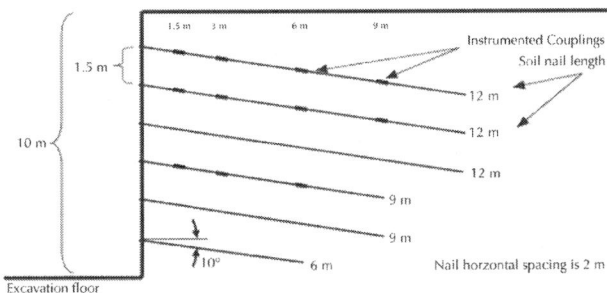

**Figure 1**: The full scale soil nail retaining structure modelled in the first centrifuge test (Jacobsz & Phalanndwa 2011).

Although the survival rate of the soil nail instrumentation was poor, it showed that the maximum axial forces in the top soil nail stabilised at approximately 50 kN, approximately two thirds of the load calculated using a simple failure wedge analysis. It was found that soil nail loads were not mobilised gradually, but in distinct load increments. It appeared that the material behind the excavation remained stable to a point as the excavation advanced and, only when a certain excavation depth was reached, did the retained soil exert more load on the soil nails, as it depended on the nails for stability. Soil nail loads were mobilised in a number of such load steps as the excavation advanced, as illustrated in Figure 2.

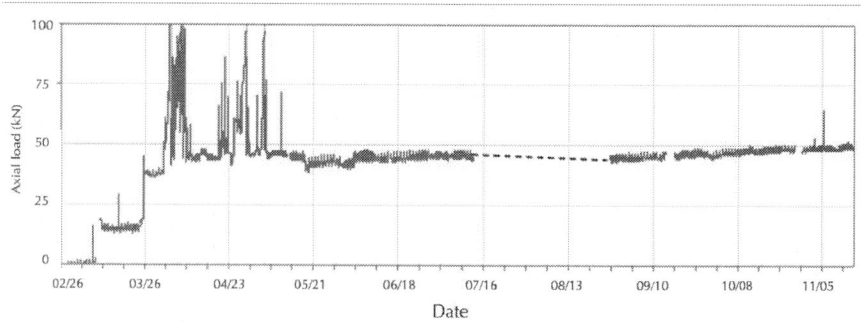

**Figure 2**: Axial load variation in the top instrumented soil nail (Jacobsz & Phalanndwa 2011).

The aims of the centrifuge model study were:

- to measure the load mobilisation in the soil nails over time during and after excavation, and
- to compare the mobilised soil nail loads in the model with those from the Jacobsz & Phalanndwa (2011) case study, and with those calculated from conventional wedge theory.

# CENTRIFUGE MODEL

A centrifuge model was set up to model the soil nail wall described in the Gautrain retaining wall case study. The model was constructed at a scale of 1:50 and was therefore tested at an acceleration of 50 G. The scale factor was chosen taking into account the dimensions of the model container, referred to as a strong-box, in relation to the dimensions of the full-scale situation being modelled. The model is illustrated diagrammatically in Figure 3.

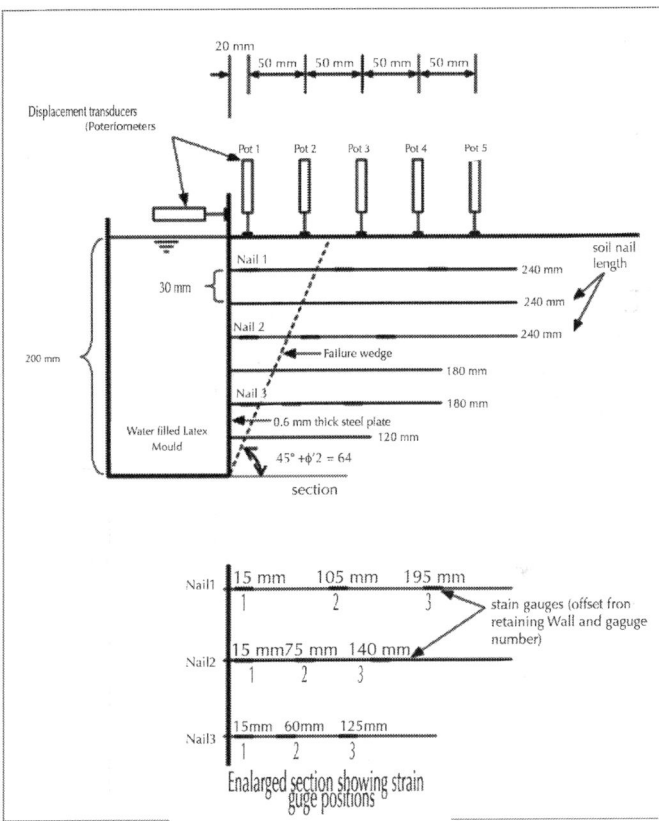

**Figure 3:** The centrifuge model (not to scale).

The model retaining wall was constructed from a 0.6 mm thick galvanised steel plate. The calculated bending stiffness (EI) of the shotcrete facing, assuming an un-cracked panel, was approximately $9.4 \times 10^6$ $Nm^2/m$ (assuming a Young's modulus for concrete of 20 GPa and 200 GPa for steel). Bending stiffness scales with the fourth power of the scale factor. The bending stiffness of the plate used to model the shotcrete face was calculated at 3.6 $Nm^2/m$, which was therefore approximately 2.4 times stiffer than the scaled-down retaining wall value.

The model soil nails were made from 5 mm wide brass strips, 0.2 mm thick, which were bolted to the wall using 2 mm diameter nuts and bolts. The reason for using flat metal strips was so that the model soil nails could easily be instrumented with strain gauges. For ease of installation during model preparation, the nails were installed horizontally.

The purpose of the model was to investigate the mobilisation of axial loads along the length of the nails during excavation, i.e. to simulate normal operational conditions and not to fail the soil nail wall. Disregarding the effects of dilation, the design pull-out capacity of the soil nails, calculated purely from interface friction between the nails and the soil, therefore exceeded the imposed load estimated from active pressure on the wall by approximately one third, providing a safety margin. The pull-out load ($Q_u$) of the flat strip model soil nails was calculated from $\sigma_v A_n \tan\delta$, where $\sigma_v$ is the vertical stress acting at the depth of the nail, $A_n$ the surface area of the nail (top and bottom) and $\delta$ the interface friction angle between the sand and the brass strips, measured in a shear box test at 26°. A total pull-out force of 1272 kN (full-scale) was calculated for a column of six nails. The predicted active pressures to be resisted per column of nails were 932 kN.

The calculated axial stiffness of the full-scale nails is approximately 100 MN. Axial stiffness scales with the square of the scale factor. The required stiffness of the model nails was therefore 40 kN. The brass strips were 2.7 times stiffer than the scaled requirement. It was, however, not practical to use narrower strips due to instrumentation difficulties.

Three model nails were instrumented with three strain gauges each, connected in quarter Wheatstone bridge circuits. The strain gauges were positioned with the first gauge close to the wall and the second gauge close to the position where the maximum tensile force was expected, i.e. where an active failure wedge is expected to be mobilised (roughly at an angle of 45° + $\phi$/2 with the horizontal) (e.g. Lazarte et al 2003). The third gauge was mounted approximately halfway between the second gauge and the end of the soil nail (see Figure 3).

The soil used in the model was a fine alluvial silica sand sourced from a commercial source near Cullinan. It was found that particles larger than approximately 200 µm were relatively well rounded, but the finer fraction tended to be more angular with a description of angular to sub-angular being appropriate. The grading curve for the sand is presented in Figure 4. The friction angle of the sand was measured at 37° using a conventional shear box. During model preparation the sand was placed by pluviation during which a constant drop height and flow rate were maintained. The sand was pluviated in layers of about 30 mm thickness, i.e. the vertical spacing between rows of soil nails. The placed relative density of the sand was approximately 55% (1566 kg/m³), i.e. a medium dense sand. The mass of sand was determined by weighing the model before and after placing the sand.

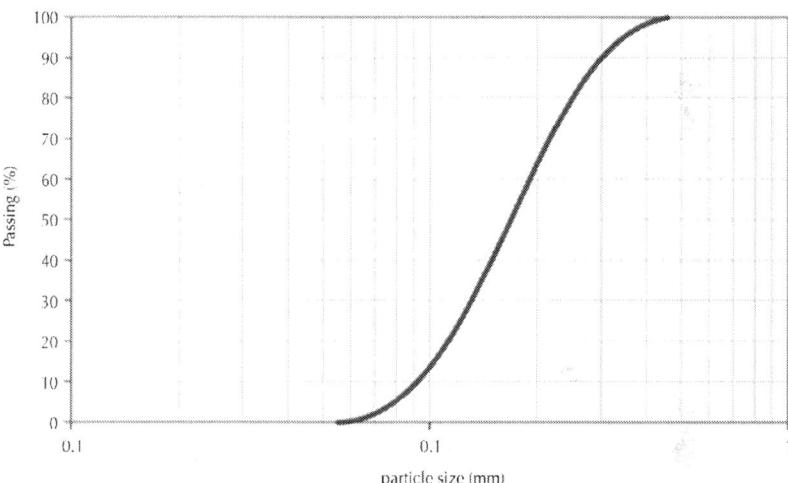

**Figure 4:** Sand grading.

The deepening of the excavation was modelled using a water-filled Latex rubber mould in which the water level was reduced during the test. This method was also used by Tei (1993) (see also Tei *et al* 1998). During the acceleration of the centrifuge to 50 G, the water level in the rubber mould was maintained at the correct level using a standpipe with a fixed overflow level into which water was continuously fed. This procedure was followed because it was expected that during acceleration of the centrifuge some movement of the system would have occurred, possibly affecting the water level in the rubber mould which would disturb the stress regime. After accelerating to 50 G, the water supply to the standpipe and rubber mould was stopped. A solenoid valve was opened to release the water from the rubber mould to model the excavation of soil in front of the retained face. In the first test the water level was allowed to drop without interruption from 200 mm to 0 mm depth. In the second test the water level reduction took place in steps over 2 000 seconds, and in the final test over 3 000 seconds. After every step in water level reduction, some horizontal wall movement took place, which took some time to stabilise. The next drop in water level was only initiated after this wall movement had stabilised.

During the tests the vertical movement of the sand surface and the horizontal movement at the top and mid-height of the retaining wall were monitored using potentiometer-based displacement transducers. The water level in the rubber mould was monitored using a pressure transducer mounted near the base of the standpipe. A number of photos of the model are presented in Figure 5.

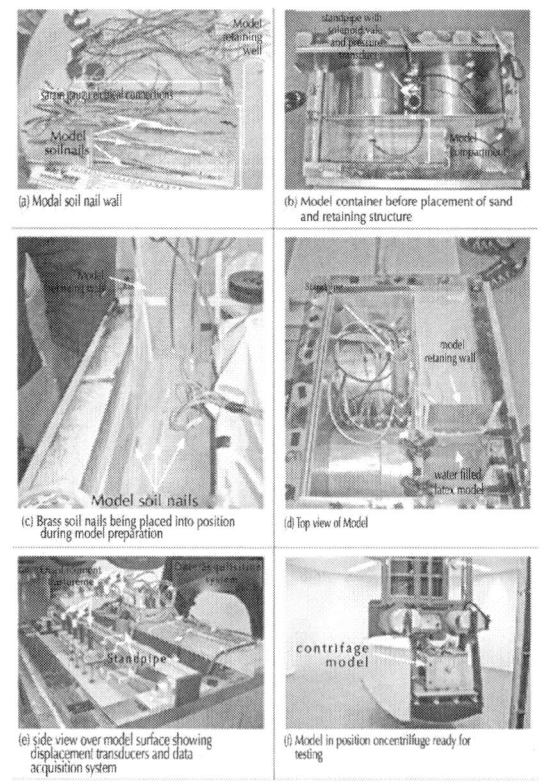

(a) Modal soil nail wall

(b) Model container before placement of sand and retaining structure

(c) Brass soil nails being placed into position during model preparation

(d) Top view of Model

(e) side view over model surface showing displacement transducers and data acquisition system

(f) Model in position on centrifuge ready for testing

**Figure 5:** Sequence of photos illustrating model preparation.

# CENTRIFUGE MODEL TEST RESULTS

## Surface Settlement

Surface settlements were recorded with potentiometers with a resolution of approximately 0.001 mm during the lowering of the water level. During the acceleration of the centrifuge to 50 G the upper surface of the sand settled between 1 mm and 2 mm in response to the stress increase acting on the model. Once at 50 G, the settlement data was zeroed so that the surface settlements caused by the lowering of the water level behind the retaining wall could be measured. Figure 6 shows the settlement of the soil surface behind the retaining wall in response to the lowering of the water level.

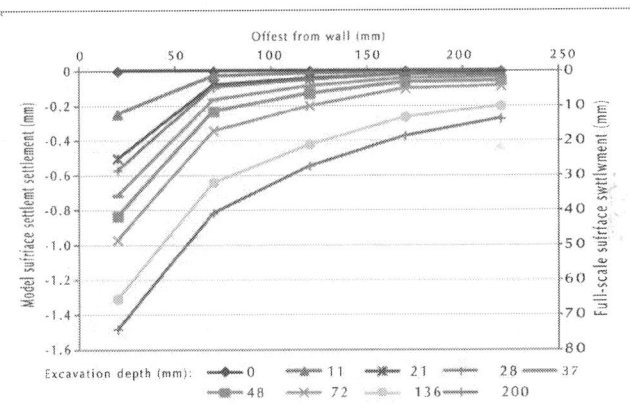

**Figure 6:** Surface settlement in response to "excavation".

A maximum settlement of approximately 1.5 mm occurred immediately (20 mm) behind the wall and reduced with distance away from the wall. This translates to 75 mm at the full scale (1:50).

## Horizontal Wall Movement

Figure 7 presents the horizontal movement measured at the top and mid-height of the retaining wall in response to lowering of the water level, modelling the excavation. The results of the three tests show good repeatability between tests and illustrate that the rate of water level reduction did not have a significant effect on the wall movement.

It can be seen from the figure that as the water level began to be lowered, wall movement immediately began to occur at the top of the wall. When the water level in the model excavation had dropped to below the depth of the first row of soil nails (30 mm), the rate of movement decreased as the nails began to restrain wall movement. The rate of wall movement then remained approximately constant as the excavation advanced.

Little horizontal movement was observed at the mid-height position on the wall until the water level had reduced to that height. Thereafter, horizontal movement occurred at approximately the same rate as the horizontal movement at the top of the wall.

Once the model excavation had been emptied completely, a maximum horizontal movement of about 2.5 mm was observed at the top of the wall, equating to 125 mm for the full-scale wall. The wall remained stable after excavation.

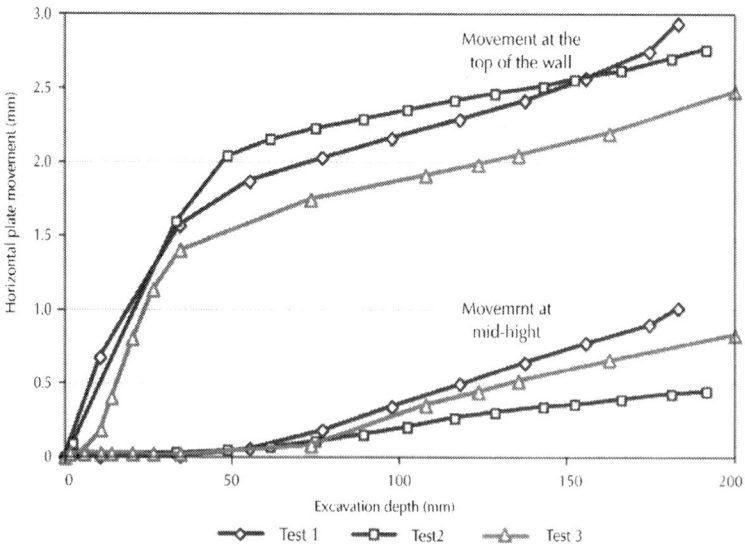

**Figure 7:** Horizontal wall movement in response to increasing excavation depth.

# Mobilisation of Soil Nail Forces

The development of axial loads in the soil nails in response to the deepening excavation is presented in Figure 8. During acceleration of the centrifuge to 50 G some settlement of the model wall relative to the sand occurred so that the parts of the nails close to the wall were subjected to a small amount of bending. This affected the zero offsets of force readings registered by the instrumented nails. Soil nail readings were therefore zeroed prior to the water level in the model excavation being reduced, to give loads mobilised due to the reduction in the water level only. Loads prior to zeroing were generally small (less than 10 N at model scale), except where bending of the nails occurred. The loads measured in the model are shown on the left-hand axis, with full-scale (prototype) loads on the right-hand axis. The calculated loads for the model from the wedge analysis based on friction angles of 30° and 37° are also shown in Figure 8; the comparison is discussed later.

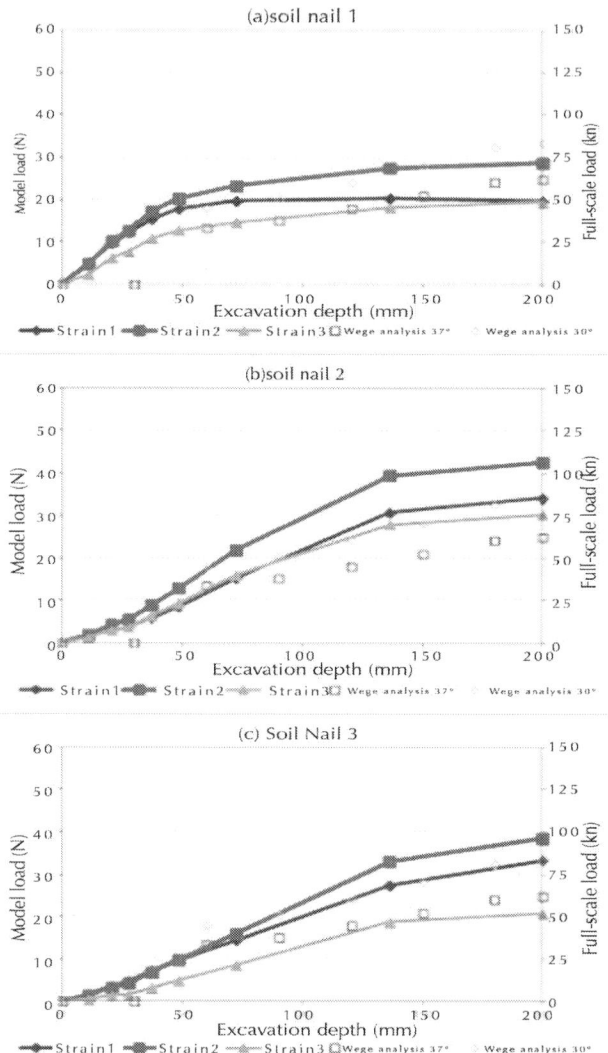

**Figure 8**: Development of soil nail forces with increasing excavation depth.

The evolving axial load distributions in the instrumented nails, as the excavation was deepened, are presented in Figure 9. Initially, the highest loads were mobilised immediately behind the wall in response to active pressure behind the wall, but soon the location of maximum force migrated backwards from the wall as a failure mechanism began to mobilise.

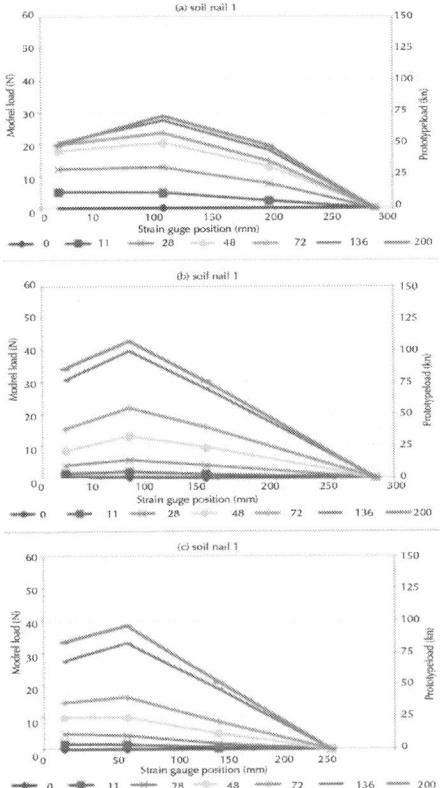

**Figure 9:** The distribution of soil nail forces along their lengths as excavation depth increases.

# DISCUSSION

## Comparison of Model Results with Analytical Methods

### *Wedge Analysis*

The equilibrium of a simple triangular active failure wedge behind the excavation face was examined to estimate the development of axial soil nail forces in response to the deepening excavation (Figure 10). This approach is commonly used for soil nail design, although the complexity of the mechanisms varies (SAICE 1989). For the problem modelled in the centrifuge, only three forces were considered: the self-weight of the failure wedge (W), the resisting force mobilised on the failure plane (R) and the sum of the individual soil nail forces (T). For a fully mobilised failure mechanism the resisting force R would act at an angle φ as shown inFigure 10, where φ is the soil friction angle. The soil nails were assumed to carry only axial loads, disregarding any bending or shear stiffness they might possess. The failure wedge was assumed to mobilise at a slope angleβ. This slope angle was varied to find the maximum axial soil nail force (T). For a horizontal soil surface and smooth vertical retaining wall, the wedge analysis provides the same solution as the active Rankine earth pressure case.

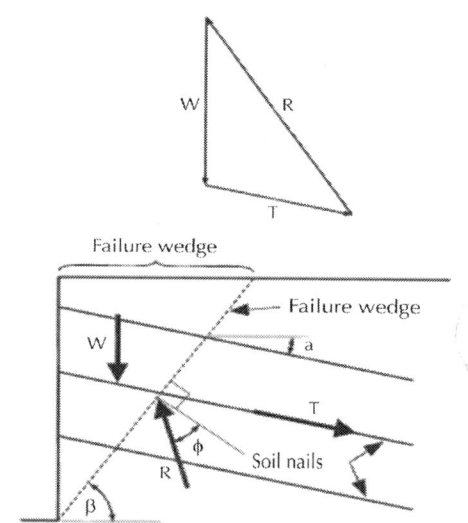

**Figure 10**: Simplified wedge analysis used for the estimation of soil nail forces.

The soil nail loads were calculated for various depths of excavation by simply dividing the total calculated soil nail force (T) by the number of nails intersecting the failure wedge. The calculated forces (based on horizontal soil nails) are plotted with the observed loads in Figure 8. As no failure wedge intersects soil nails for excavation depths of up to

30 mm (1.5 m at prototype scale), zero soil nail force was assumed up to this depth.

## Soil Nail Forces

Figure 8 illustrates that the loads in the soil nails initially increased approximately linearly with increasing excavation depth, but the rate of increase reduced with further excavation.

The trend in the measured soil nail forces compares well with the predictions from the wedge analysis, although the latter generally tends to underestimate the loads. This is somewhat in contrast with Shen *et al* (1982), Tei *et al* (1998), Lazarte *et al* (2003) and others who stated that average nail forces are generally smaller than those calculated by considering full active earth pressures. The most significant underestimation occurred on the second soil nail.

During the acceleration of the centrifuge to 50 G it was attempted to balance the earth pressures behind the model retaining wall by maintaining a constant water level in the rubber mould as described. However, some vertical and horizontal movements of the various components of the model were unavoidable during acceleration. The imperfect method of balancing the earth pressures as described, in combination with the movements that occurred during acceleration, resulted in a certain amount of load mobilising in the soil nails prior to reducing the water level in the rubber mould to model excavation. This means that a portion of the shear strength of the sand was already mobilised prior to water level being reduced. Because of zeroing of the soil nail reading prior to reducing the water level, these loads were ignored. The various disturbances would most probably have resulted in the amount of shear strength mobilisation in the sand before excavation to be different from the situation applicable to an actual soil nail wall, probably resulting in less shear strength being available to support the excavated face than what would have been expected. The implication of this is that the soil friction angle used in analysing the model should probably be reduced. When a friction angle of 30° is used instead of 37°, the correlation between the measured soil nail forces and those calculated using a wedge analysis improves (see Figure 8).

A further factor contributing to the difference between the measured and calculated loads is the fact that the actual stress distribution behind the retained face is significantly more complex than the simple triangular distribution assumed by active earth pressure theory (Tei 1993 and Tei *et al* 1998). Tei (1993) states that the failure surfaces in sand would resemble a logarithmic spiral which would result in failure wedges that are approximately 10% heavier than the assumed triangular wedge. Also, Zhang *et al* (2001) mentioned that the failure wedge in the presence of soil nails was deeper than without reinforcement. The actual mobilised soil nail forces are controlled by many factors, including the flexibility of the facing wall and soil nails and dilation on the soil-nail interface (Tei *et al* 1998).

The magnitude of the scaled-up maximum observed soil nail forces in the centrifuge model are put into context by comparison with normalised soil nail forces measured at eleven sites presented in Figure 11 (Byrne *et al* 1998). Observed maximum tensile nail forces were normalised by $K_a H g S_h S_v$, where $K_a$ is the coefficient of active earth pressure, H the wall height, $\gamma$ the density of the retained material and $S_h$ and $S_v$ the respective horizontal and vertical nail spacing. The figure shows that the general trend is for soil nail forces to reduce somewhat with depth, but very significant scatter occurs, probably as a result of variations in soil strength and stiffness between sites which were not taken into account in the normalisation. The observations from the centrifuge tests plot well within the data set presented in the figure.

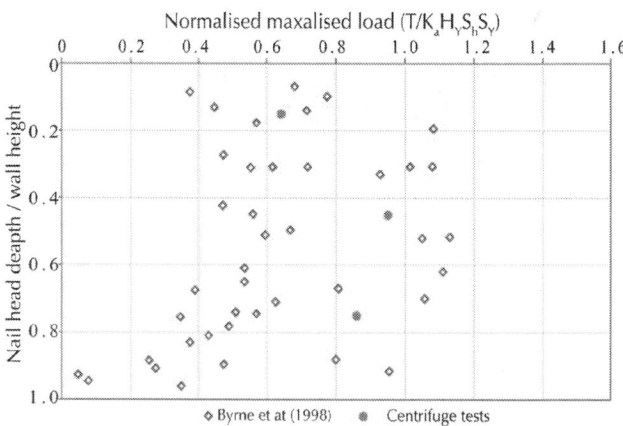

**Figure 11:** Normalised maximum tensile forces measured in soil nail retaining walls (Byrne et al 1998).

In the Jacobsz & Phalanndwa (2011) case study, soil nail loads of just less than 50 kN were measured in the top soil nail when the system was at equilibrium. These are of the same order of magnitude, albeit somewhat lower than scaled loads from the model (see Figure 8). They are also lower than the prediction from a wedge analysis. Note that a wedge analysis predicts soil nail forces that are 12% higher when nails are installed at 10° compared to horizontal nails. The reason for the scaled model loads being higher can be ascribed to the fact that the model soil profile comprised cohesionless sand in which some shear strength had already been mobilised during acceleration of the centrifuge, while the profile in the field comprised residual andesite, possessing significant cohesive strength, increasing with depth.

A further difference between the model and the case study is the step-wise way in which loads were mobilised in the case study compared to a more gradual increase in load in the model (compare Figure 2 with Figure 8). The reason for the step-wise load increase was attributed to the fact that the excavation could support itself to a certain depth and then suddenly yielded, mobilising load in the soil nails. With further excavation, it again remained stable to a certain depth before yielding again, applying another step-wise load increase on the soil nails. The cohesion-less sand did not possess any strength to support any depth of excavation, so that axial load had to be mobilised in the soil nails very shortly after the water level in the model excavation began to reduce.

The measured axial force distributions along the length of the nails shown in Figure 9 generally agreed with the pattern typically observed in the field. A soil nail normally carries a load at the retained face which increases towards the intersection with the failure plane and then reduces to zero at the end of the nail (Lazarte et al 2003). The maximum load was measured consistently at the second strain gauge on each nail. They were purposefully installed close to where the failure wedge was expected to intersect the soil nails.

## Wall and Ground Movements

The vertical soil settlement behind the wall amounted to approximately double the amount of the expected settlement given by the guideline of H/333 by Lazarte et al (2003) for fine grained soils. However, the observed settlement applies to a medium dense sand, the material used in the model in which some shear strength had already been mobilised during centrifuge acceleration. The maximum settlement of the full-scale wall amounted to only 8 mm, illustrating that, as expected, the residual andesite behaved much stiffer than the sand in the model, settling less. The residual andesite appears to mobilise its strength at smaller strains than cohesionless sand.

It is interesting to note that the settlements above the active wedge, potentiometers 1 and 2 (see Figure 3 andFigure 6) settled significantly more than the potentiometers further away, reflecting the mobilisation of the failure mechanism. An active failure wedge is predicted to intersect the sand surface at an offset of 100 mm from the retained face. The zone behind the wall where noticeable settlements occurred, agrees well with the 140 mm (at model scale) predicted by Lazarte et al (2003).

The horizontal wall movements are presented in Figure 7 and were recorded from the onset of water level reduction until the model excavation was complete. The largest portion of horizontal movement took place during the initial reduction in water level to the depth of the first row of nails. Thereafter the rate of movement slowed considerably. In practice this initial movement would not have been recorded, because the first shot-crete panels still had to be constructed. The horizontal movement that would be recorded in practice corresponds to that associated with a drop in water level from 30 mm to the bottom of the excavation. In the tests reported here, this movement amounted to approximately 1 mm, or 50 mm at full scale.

As in the case of the vertical movement behind the wall, this horizontal wall movement also exceeded the guideline recommended by Lazarte et al (2003) (also H/333, or 30 mm at full scale). The maximum horizontal movement observed at the top of the full-scale wall was 34 mm (Jacobsz & Phalanndwa 2011). The difference can be explained due to the model comprising medium dense sand in which some shear strength had already been mobilised during centrifuge acceleration, while the full-scale situation comprised stiff residual andesite mobilising strength at smaller strains.

Following each drop in the water level in front of the model wall, it took some time before the horizontal wall movement stopped. This was also seen in the field, where some movement continued to occur for some time after completion of the excavation (Jacobsz & Phalanndwa 2011).

Figure 7 illustrates that the top of the wall deflected rapidly initially, but when the water level reached the level of the first row of soil nails, the rate of horizontal movement reduced due to the restraining effect of the soil nails. Virtually no horizontal movement took place at mid-height initially, indicating that the upper part of the wall bent above the excavation level. Once the water level reached mid-height, horizontal movement there took place at approximately the same rate as at the top of the wall, indicating that the wall translated horizontally with little further bending. This suggests that horizontal wall deformation occurred as indicated in Figure 12, with bending taking place at the excavation level while the upper part of the wall remains approximately planar.

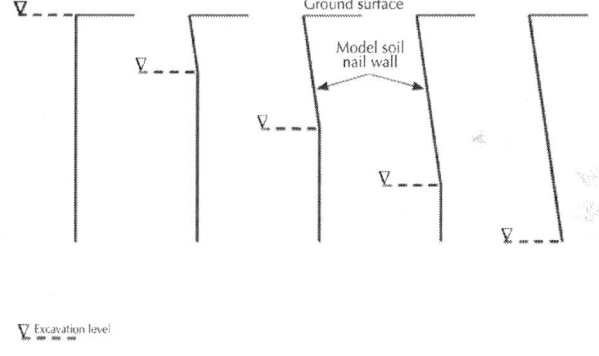

**Figure 12:** Mode of horizontal deformation of model soil nail wall in centrifuge models.

# Comparison between the Full-scale Situation and the Model

## *Soil*

It is often questioned whether the particle sizes of material used in a centrifuge model need to be scaled. For example, could the fine sand at model-scale therefore hypothetically behave as a gravel at the full-scale? In practice it is common with a centrifuge model to model the actual material occurring in the field, or often, to use the actual material from the field directly in the model. The material is then viewed as a continuum with the same stress-strain properties as in the field. Whether this assumption is reasonable depends on the ratio between the particle size in the model and the size of significant components in the model, e.g. particle size versus the dimensions of model piles, foundations or model soil nails (Taylor 1995). A method that is often used to investigate whether unrealistic scale effects occur is the so-called method of "modelling of models". Models are tested at different scales. If the scaled observations from different scale models are consistent, particle size effects can be ignored and the material can be assumed to behave as a continuum at the accelerations tested. However, when failure mechanism bounded by shear bands begin to dominate, the ratios between shear band widths, particle size and model element dimensions can become important. In such instances dilation effects within shear bands are likely to scale-up unrealistically (Taylor 1995). Milligan & Tei (1998) mentioned that relative size effects between model soil nail diameter and particle size may tend to increase the apparent strength and stiffness of the model compared to the prototype in the case of rough nails. This scale effect is significant where the ratio $D/D_{50}$ ranges from 1 to 35 (where D is the nail diameter and $D_{50}$ the main particle size), but reduces at higher values applicable in the field. Due to the thickness of the brass strips (model soil nails) relative to the means particle size, scale effects would be expected in the model. However, due to the smoothness of the model nails, dilation effects as described above should have been limited, although probably not insignificant.

## *Soil Nails*

One important aspect in which the soil nail retaining wall in the centrifuge differed from the full-scale situation was that the wall and soil nails were pre-installed prior to modelling of the excavation. Installation of soil nails during a test would be difficult. Due to the nails being pre-installed, loads could mobilise before the excavation depth had advanced to the depth of a particular row of nails. Also, installation-induced soil nail loads and soil stresses could not be modelled. These are likely to differ from the situation in the model (Milligan & Tei 1998).

# CONCLUSIONS

A physical model, examining an instrumented soil nail retaining structure, was tested successfully in three centrifuge tests. The test yielded realistic and repeatable data, comparing well with measurements made in a full-scale case study in Pretoria (Jacobsz & Phalanndwa 2011) and with a database of eleven other case studies (Byrne *et al* 1998).

In terms of soil nail forces, the model showed somewhat higher nail forces compared to those predicted by a simple equilibrium analysis and when compared with the case study discussed. This is likely to be a consequence more of the shear strength of the soil being mobilised during acceleration of the model than what would be applicable in a full-scale (K0) situation, resulting in less strength being available to resist excavation-induced loads than what would have been expected. Information from the literature suggests that soil nail forces from a simple wedge analysis or limit equilibrium analysis are conservative. The results of these centrifuge tests suggest that soil nail forces from centrifuge tests may be even more conservative, due to the mobilisation of some soil strength during centrifuge acceleration.

The axial load distributions measured along the length of the soil nails compared well with the known distributions from the literature.

The trend in axial load mobilisation in the soil nails differed from the full-scale case study reported. In the model, axial load was mobilised gradually in response to excavation, while in the full-scale field study a stepwise mobilisation was observed. The reason for this is that the soil in the model only possessed frictional strength, while the residual andesite in the field had some "cohesive" strength and a fissured structure, enabling it to remain stable up to a certain depth.

Although differences between the full scale situation and a model are unavoidable, physical modelling in the geotechnical centrifuge is a valuable technique to model complex three-dimensional problems. An advantage is that a physical event can be observed and realistic results obtained using the same materials as in the field.

# REFERENCES

1.  Bolton, M D & Pang, P R L 1982. Collapse limit state of reinforced earth retaining walls. *Geotechnique* 32(4): 349-367.

2.  Byrne, R J, Cotton, D, Porterfield, J, Wolschlag, C & Ueblacker, G 1998. *Manual for Design and Construction Monitoring of Soil Nail Walls*. Report No FHWA-SA-96-69R, Washington DC: US Federal Highway Administration.

3.  Jacobsz, S W & Phalanndwa, T S 2011. Observed axial loads in soil nails. *Proceedings,* 15th African Regional Conference on Soil Mechanics and Geotechnical Engineering, Maputo, Mozambique, 221-227.

4.  Lazarte, C A, Elias, V, Espinoza, R D and Sabatini, P J 2003. Soil nail walls. *Geotechnical Engineering Circular No. 7,* Report No FHWA0-IF-03-017, Washington DC: US Federal Highway Administration.

5.  Milligan, G W E and Tei, K 1998. The pull-out resistance of model soil nails. *Soils and Foundations,* 38(2):179-190.

6.  SAICE (South African Institution of Civil Engineering) 1989. *Lateral Support in Surface Excavations, Code of Practice.* Johannesburg: SAICE Geotechnical Division.

7.  Shen, C K, Kim, Y S, Bang, S & Mitchell, J F 1982. Centrifuge modelling of lateral earth support. *Journal of Geotechnical Engineering Division, ASCE,* 108(GT9): 1150-1164.

8.  Taylor, R N 1995. *Geotechnical Centrifuge Technology.* London: Blackie Academic & Professional.Tei, K 1993. A study of soil nailing in sand. PhD Thesis, Oxford: University of Oxford.

9.  Tei, K, Taylor, R N & Milligan, W E 1998. Centrifuge model tests of nailed soil slopes. *Soils and Foundations,*38(2): 165-177.

10. Zhang, J, Pu, J, Zhang, M & Qui, T 2001. Model tests by centrifuge of soil nail reinforcements. *Journal of Testing and Evaluation* 29(4): 315-328.

# Chapter 3

# Centrifuge Modelling for Evaluation of Seismic Behaviour of Stone Masonry Structure

Heon-Joon Park[a] and Dong-Soo Kim[b]

[a]Schofield Centre, University of Cambridge, High Cross, Madingley Road, Cambridge CB3 0EL, UK

[b]Department of Civil and Environmental Engineering, KAIST, 291 Daehak-ro, Yuseong-gu, Daejeon, 305-701, Republic of Korea

## ABSTRACT

Many surviving ancient monuments are freestanding stone masonry structures, which appear to be vulnerable to horizontal dynamic loads such as earthquakes. However, such structures have stood for thousands of years despite numerous historic earthquakes. This study proposes a scaled-down dynamic centrifuge modelling test to study how these masonry structures resist seismic loading. The test is proposed for seismic risk assessments to evaluate risk of damage from a future seismic event. The seismic behaviour of a 3-storey, freestanding stone block structure has been modelled and tested within a centrifuge. Models were made at 3 different scales and dynamic tests were conducted using different centrifugal acceleration fields so that the behaviours could be transformed to an equivalent full-scale prototype and compared. Data from 2 earthquakes and a sweeping signal were used to simulate the effects of earthquake ground motion within the centrifuge. The acceleration and frequency responses at each storey height of the model were recorded in different centrifugal acceleration fields. Similar behaviours appeared when the results of the small-scale models were transformed to a full-size prototype scale. This confirms that the seismic behaviour of stone masonry structures can be predicted using scaled-down models.

# INTRODUCTION

Heritage monuments constructed using stone masonry appear to be vulnerable to horizontal dynamic loads such as earthquakes. Culturally significant stone structures, such as freestanding columns, however, have stood for over a thousand years despite numerous historic earthquakes. This fact leads to 2 questions: how are these structures able to survive multiple earthquakes, and how safe are they for future seismic events? There have been many studies on the dynamic behaviour and seismic vulnerability of stone masonry structures. However, analytical approaches and numerical modelling are difficult and the results are often inconclusive. A full-scale shake table test yields the best data, but such a test is costly. Because friction governs the behaviour of stone masonry structures, reduced-scale centrifuge modelling can be an effective alternative.

The stress conditions of a scaled model in a centrifuge test are known to be the same as in a full-scale prototype. Thus, this equivalent stress condition makes it possible to simulate real seismic behaviour using a reduced-scale model. This study proposes a dynamic centrifuge test for structural analysis and seismic risk assessment as a means to evaluate its safety for future earthquakes.

In this study, the scaling factors are examined through a 'modelling of models' procedure, which is used to evaluate the effects due to scaling. Scaled testing is particularly useful when no full-scale test data is available [1]. Models of 3 different scales were made, representing a 3-storey rectangular parallelepiped stone structure. The dynamic response of each block was recorded using accelerometers installed at the middle of each block. Acceleration data was collected at different centrifugal acceleration fields (g-levels). Two real earthquakes, Hachinohe and Ofunato, and a sine-sweeping signal were used as input accelerations in the centrifuge. The tests were conducted at different g-levels so that the recorded behaviours of one model scale at different g-levels could be transformed to simulate an imaginary full-scale test and compared.

# DYNAMIC BEHAVIOUR OF STONE MASONRY STRUCTURE

The cyclic or dynamic behaviours of block structures, such as ancient stone monuments freestanding on their foundations, have been subject to numerous studies. Cyclic displacements due to the elastic response of the soil–foundation system under vibration loading have been the primary focus. Theoretical solutions for rocking and sliding motions of rigid rectangular foundations have been proposed [2], [3], [4], [5] and [6]. Since the pioneering research of Housner, various studies have been conducted on the dynamic behaviour of single block structures in rectangular shapes. Housner produced a basic understanding of the rocking behaviour of a block and derived a useful mathematical model [7]. Ishiyama [8] classified the behaviours of single blocks into 6 categories and established governing equations for each category, through which he studied the natural characteristics and mode conversion standards of each category. Spanos and Koh [9] have studied the steady state modes of a rigid body by analysing its behaviour in harmonic motion. Also, Tso and Wong [10] approached these modes experimentally.

In the case of multi-block systems, determining the governing equations and mode conversion of the entire system becomes analytically difficult because of the complicated boundary conditions between blocks. Psycharis [11] proposed a governing equation, assuming that rocking takes place only in 2 slender block structures. Makris and Zhang [12] examined the transient rocking response of anchored blocks subjected to horizontal pulse-type motion. Spanos et al. [13] systematically reviewed various analytical, probabilistic, and experimental assessment studies of dynamic behaviour of block structures, including the above-mentioned studies. Makris and Konstantinidis [14] examined the distinct characteristics of the rocking spectrum for a slender rigid block. Since this study, the dynamic behaviours of 2 rigid block structures caused by ground motion have been organised analytically according to patterns and numerically compared.

Today, complex numerical analysis of multi-block systems is possible. And shake table tests and seismic risk assessments are routinely conducted for significant historic stone monuments. Psycharis et al. [15] and [16] and Konstantinidis and Makris [17] presented a numerical investigation on the seismic response for multi-drum classical columns of ancient monuments. Kim and Ryu [18] produced a full-scale model of a stone pagoda in Sang-Gye-Sa Temple that had been damaged by an earthquake in 1936. They conducted a 1g shake table test to predict the accelerations in the monument caused by the earthquake. D'Ayala et al. [19] analysed a series of shake table tests on 3 1/10-scale 3D dry masonry models. Peña et al.[20] described the dynamical behaviour of various freestanding block structures under seismic loading using 1g shake table tests. Konstantinidis and Makris [21] also investigated the seismic response of 1/4-scale models of freestanding laboratory equipment subjected to strong earthquake shaking through experimental and analytical studies. Kounadis et al. [22] discussed the dynamic stability rocking response of such a rigid block, 2-DOF systems subjected to horizontal harmonic ground motion using a closed form solution and numerical study. D'Ayala and Ansal [23] authored the guidelines 'Risk assessment of cultural heritage buildings' to address the vulnerability of cultural assets, specifically buildings with international cultural value.

The dynamic behaviour of a stone monument during a seismic event includes complex behaviours combined by sliding and rocking motions. In the case of multi-block systems, there are many analytical and numerical difficulties when trying to predict and understand the various mode characteristics of the materials and the interaction between units. Currently, a full-scale 1g shake table test is the most useful method for acquiring useful data on structural behaviour, but such a test is time and resource intensive.

This study proposes a dynamic centrifuge test to evaluate the seismic behaviours of stone structures for seismic risk assessment of heritage monuments. For multi-block systems, the friction characteristics between stone units have a significant effect on the behaviour of the structure. The centrifuge test described in this paper simulates the $N$ times gravity accelerations for a 1/$N$ scaled model. If an appropriate scaling law is used and the model material has the same density as that of the real structure, the seismic behaviour can be observed under the same stress conditions as those of reality. Therefore, this method is a useful alternative to numerical analysis and full-scale shake table tests. With this said, there exists a limitation in this method in that it is difficult to reproduce the same surface roughness between modelled stone structure and a full-scale prototype.

Scaling laws are the main factors in testing a full-size prototype as well as a small-scale model. The laws used for the centrifuge test are listed in the second column of Table 1[1]. In the case of a 1g shake table test for the reduced model, a scaling law has been established by Iai et al. [24]. When the density and shear wave velocity of the prototype and model are same, Iai's scaling factors are summarised as in the third column of Table 1. When comparing the scaling laws of a dynamic centrifuge test and 1g shake table test for a 1/$N$ scaled model, the biggest difference is the stress condition. For the centrifuge test, the stress condition is the same as in the full-size prototype in terms of gravity direction and horizontal direction. This equivalent stress condition supports the possibility of accurately simulating the seismic behaviour of historic stone monuments in a dynamic centrifuge test on a reduced-scale model.

**Table 1**: Scaling laws for dynamic centrifuge test and 1g shake table test

| Quantities | Scaling factors (prototype/model) | |
|---|---|---|
| | Dynamic centrifuge test | 1g shake table test |
| Displacement, length | $N$ | $N$ |
| Acceleration, gravity | $N^{-1}$ | 1 |
| Mass | $N^3$ | $N^3$ |

| Density | 1 | 1 |
|---|---|---|
| Stress | 1 | $N$ |
| Strain | 1 | 1 |
| Time (dynamic) | $N$ | $N^{0.5}$ |

# EXPERIMENT SETUP FOR MODELLING OF MODELS

This study was conducted at a recently established centrifuge facility at the KOCED Geotechnical Centrifuge Center at KAIST [25] and [26]. The centrifuge is able to simulate seismic motion by spinning at a desired centrifugal acceleration. The KOCED earthquake simulator used in this research is of an electro-hydraulic-servo type with rotation radius of 5.0 m and maximum capacity of 240g. This earthquake simulator provides 40 g shaking acceleration for no payload and 20g shaking acceleration for up to 700 kg of payload. A 40g centrifuge acceleration is equivalent to a peak ground acceleration (PGA) of 0.5g. Korean seismic design has a maximum seismic design acceleration (for rock outcrop) of 0.22g.

The 'modelling of models' technique was performed on a 3-storey rectangular parallelepiped stone structure. Each storey level consisted of one block;each block has identical size, mass, and contact surface properties. Models were made at 3 different scales so that the tests could be performed at different g-levels, behaviours transformed to full-scale, and compared in order to simulate the behaviour of one full-scale prototype. Tests were conducted using 3 sets of differently scaled models in separate test sets. For example, the dynamic motions of models 1, 2, and 3 were recorded at centrifuge accelerations of 10g, 15g, and 20g, respectively. These 3 test results, all at the same model scale, were aggregated as prototype 1. This was labelled as Test Set 1. Table 2 summarises the entire experiment setup.

**Table 2.**Experimental setup for the modelling of models

|  |  | Model 1 | Model 2 | Model 3 | Prototype size ($W$ m×$D$ m×$H$ m) |
|---|---|---|---|---|---|
| Size of one block ($W$ cm×$D$ cm×$H$ cm) |  | 36×18×9 | 24×12×6 | 18×9×4.5 |  |
| Mass of one block (kg) |  | 15.6 | 4.61 | 1.95 |  |
| Story (the number of blocks) |  | 3 | 3 | 3 |  |
| Centrifugal acceleration (g-level) | Test Set 1 (prototype 1) | 10g | 15g | 20g | 3.6×1.8×0.9 |
|  | Test Set 2 (prototype 2) | 20g | 30g | 40g | 7.2×3.6×1.8 |
|  | Test Set 3 (prototype 3) | 30g | 45g | 60g | 10.8×5.4×2.7 |

The stone used in the experiment is Hwang-deung stone, a type of homogenous granite. All the contact surfaces were cut using a water jet. The density of the stone specimen is 2.67 t/m$^3$ and the shear wave velocity of the specimen is 2037 m/s, obtained using a free–free resonant column (FFRC) test [27]. Two earthquakes were used as inputs: Ofunato earthquake with high short-period accelerations and the Hachinohe earthquake with high long-period accelerations. In addition, a sine-sweeping signal within the shaking frequency range was used, as shown in Fig. 1. The excitation inputs were applied in the transverse and longitudinal directions. 3–4 shakings were conducted for each earthquake simulation, from the lowest acceleration to the PGA, with a return period of 2400 years (PGA of 0.15g–0.28g in prototype for each test set).

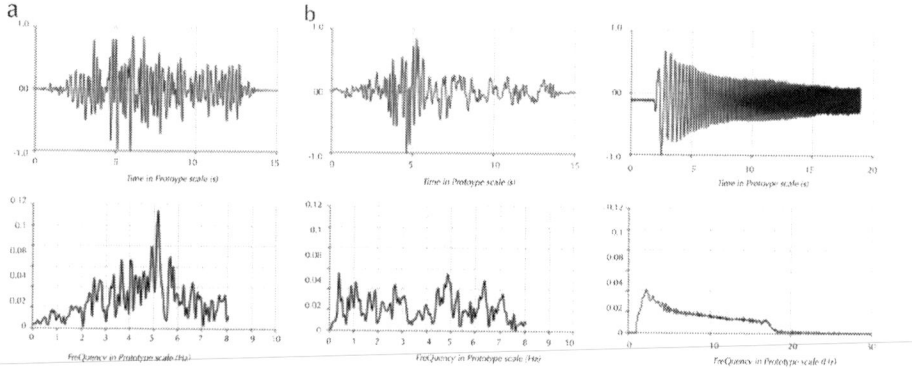

**Figure 1**: Time histories and frequency contents of input signals for shaking: (a) Ofunato earthquake record; (b) Hachinohe earthquake record; and (c) a sine-sweeping signal.

The stacked blocks of the model were built on a compacted soil layer. The acceleration of each block was recorded and analysed using the data recorded by PCB353B17 (PCB Piezotronics) accelerometers placed in the middle of each block. The experimental method for Test Set 1 and the location of the accelerometers, are shown in Fig. 2.

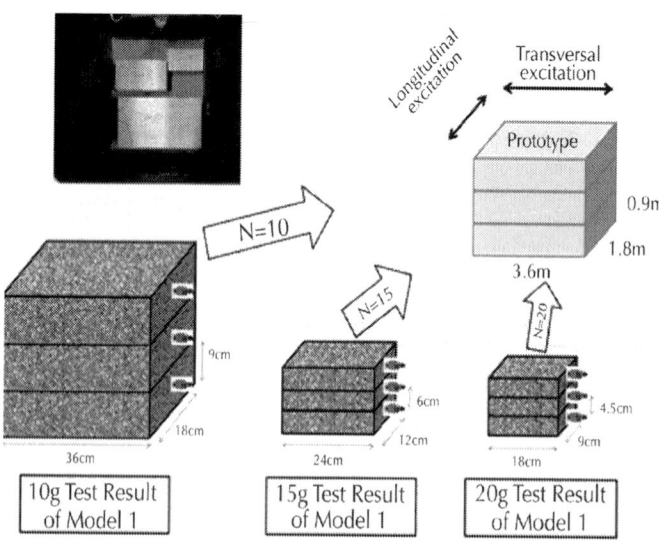

**Figure. 2**: The experiments method for Test Set 1 of modelling of models.

Dynamic centrifuge tests were conducted at several g-levels for each model scale. For each model set, the acceleration values were measured for each block and used to predict the behaviour of an imaginary full-scale, 3-storey prototype. The parameters used for comparison were peak acceleration in time domain and predominant frequency in the frequency domain.

# TEST RESULTS IN TIME DOMAIN

The procedure for comparing the experimental results of a model at a 1/N scale is as follows. First, the acceleration data were recorded at the ground surface, the first floor, second floor, and third floor. Second, these measured values were transformed into equivalent full-scale values by applying the scaling law shown in Table 1. In other words, time was applied by $N$ times, and the acceleration amplitude is applied by 1/Ntimes into the experiment results of $Ng$ state. Third, peak accelerations were determined for comparison. This was done for each model scale so that the experimental results of the 3 model types could be compared at a full-size prototype scale. Because it is difficult to keep the peak ground surface acceleration constant, the results were normalised by dividing the accelerations recorded at each block by the ground surface acceleration. Lastly, comparisons were made by the average value of each model test result.

Following the above procedure, peak acceleration values were recorded at centrifuge accelerations of 10g, 15g, and 20g, simulating the Hachinohe earthquake in the transverse direction for the first prototype in Fig. 3.Fig. 3(a) shows the accelerations recorded at each height of the model. Time domain signals were recorded at each block and the peak acceleration values were obtained. Peak acceleration values were recorded for 3 excitations at 10g and 4 excitations each at 20g and 30g. Fig. 3(b) shows the accelerations that were applied with the scaling law for the prototype structure from Fig. 3(a). The legend of the peak acceleration diagram inFig. 3(a) and (b) represents the centrifugal acceleration level and peak ground acceleration value.

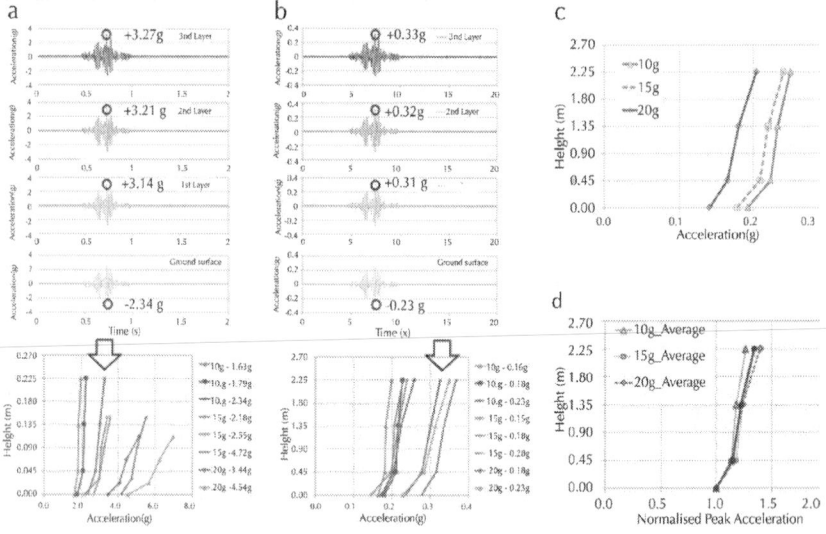

**Figure 3**: Typical test results of Test Set 1 for peak acceleration: (a) typical acceleration signal and peak acceleration values in model scale; (b) acceleration values in prototype scale; (c) averaged acceleration values; and (d) comparison of normalised averaged acceleration values.

In Fig. 3(c), average values at each g-level are shown. Finally, in Fig. 3(d), normalised average values are shown. In the case of peak acceleration values, amplification patterns at each height are similar, with a 5.7% maximum error occurring, centred on the average value. The peak acceleration values have significant differences among model 1 (10g), model 2 (15g), and model 3 (20 g), as shown in Fig. 3(a). However, after applying scaling factors and organising each peak acceleration value, a consistent amplification pattern can be seen for a given prototype, as shown in Fig. 3(d).

Fig. 4 shows the comparative graph of averaged peak acceleration values in prototype scale and the amplification patterns for all test sets. The left figures pertain to transversal excitation (long-length direction,Fig. 2), and the right figures pertain to the longitudinal direction. The maximum errors, centring on average values, are noted for each test set (1–3), each input signal (Hachinohe and Ofunato earthquake), and each shaking direction (transversal and longitudinal).

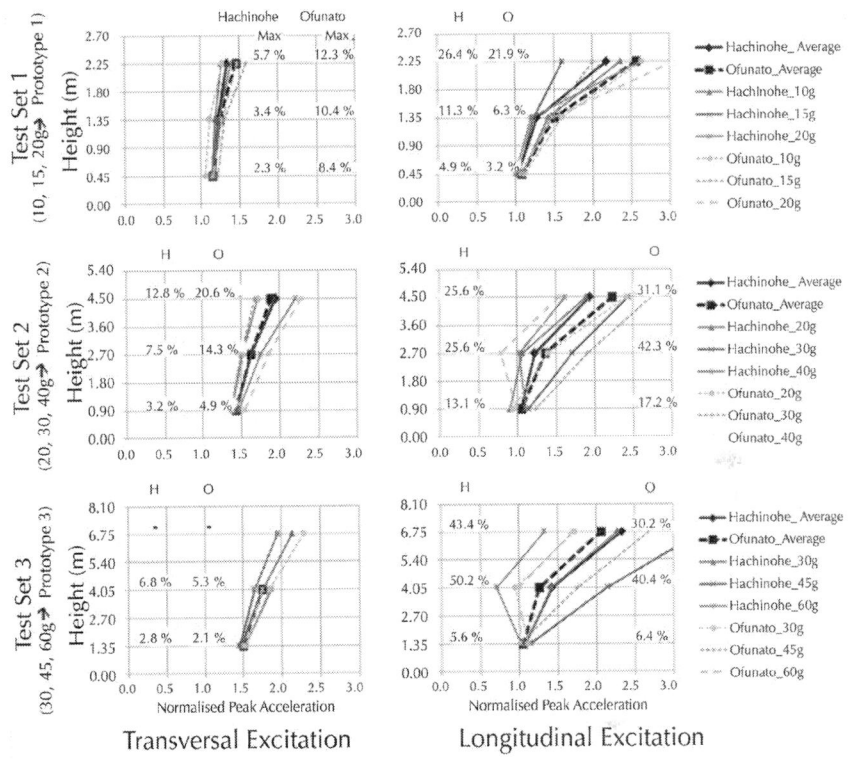

**Figure 4**: Comparison of averaged peak acceleration values in prototype scale.

When examined according to the shaking direction, the error values were higher for the longitudinal shaking compared to the transversal shaking. The error was higher in the short direction of the rectangular models compared to their long direction. In the case of transversal excitation, the amplification pattern was consistent with height. Maximum errors of 12.8% and 20.6% occurred at the third storey for the Hachinohe and Ofunato earthquake simulations, respectively. On the other hand, the amplification varied with height for longitudinal excitation. The amplification ratio between the first to second storey was different from that of the second to third storey. For 2 cases in Test Set 2 and for 2 cases in Test Set 3, the peak accelerations were less in the second storey. This can be explained by the greater complexities of displacement and rocking behaviours in taller, 3-storey masonry structures in the shaking direction.

Overall, the error values for the peak acceleration were the highest for Test Sets 2 and 3. In other words, the error was greater as the centrifuge test increased in g-level. In addition, acceleration data from the third storey block that were not measurable in the 60g state for Test Set 3 are not shown by a graph. As the g-level increased, the change in the measured results became more pronounced and the error values became greater. In most cases, however, the amplification characteristics of acceleration in the time domain could be obtained using scaling laws despite experimental errors.

# TEST RESULTS IN FREQUENCY DOMAIN

The predominant frequency was identified in the frequency domain, in which amplification takes place according to height for each setup. As with the time domain, test sets of the same scale were compared. The acceleration signals at model scale were converted to the frequency domain. They were transformed to full-size prototype scale by applying a scaling law of $1/N$ times frequency and $1/N$ times spectral acceleration. Results of shaking for the sine-sweeping signals in the frequency domain are compared, which have relatively wide and even frequency contents, as shown in Fig. 1(c).

Fig. 5 shows the frequency domain signals for transverse excitation in the prototype scale by applying the general scaling law. Fig. 6 shows the same results for longitudinal excitation. In the case of transverse excitation, the frequency contents in full-size prototype scale for 3 results of Test Set 1 are amplified, with height, at 12.6–14.8 Hz. These were derived from a 131 Hz peak in model 1 (10g), a 222 Hz peak in model 2 (15g), and a 253 Hz peak in model 3 (20g). The predominant frequencies in model scale have large differences among the 3 models. However, after applying scaling factors, a consistent predominant frequency range can be obtained for a given prototype. The frequency contents at 6.8–7.6 Hz are amplified in Test Set 2 and the frequency contents at 4.5–6.8 Hz are amplified in Test Set 3. Similar to the time domain analysis, the predominant frequency ranges were wider for longitudinal shaking, as shown in Fig. 6. For the majority of test results, it was found that acceleration amplification took place and the frequency ranges generally coincided.

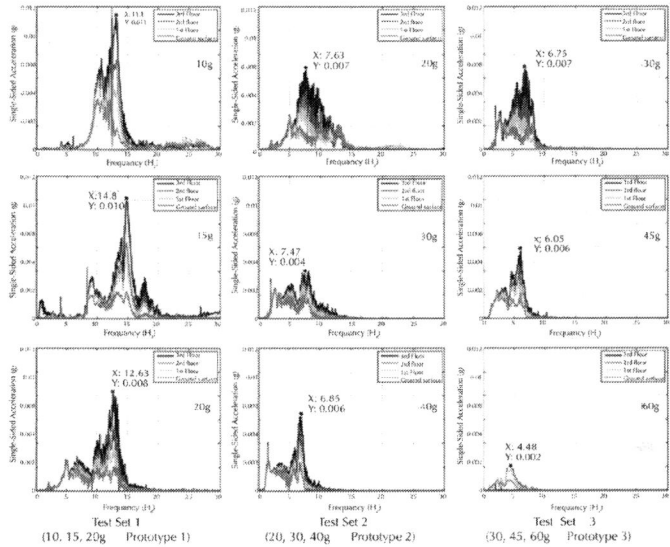

**Figure 5**:Frequency domain signals obtained from each test set for transversal excitation in the prototype scale.

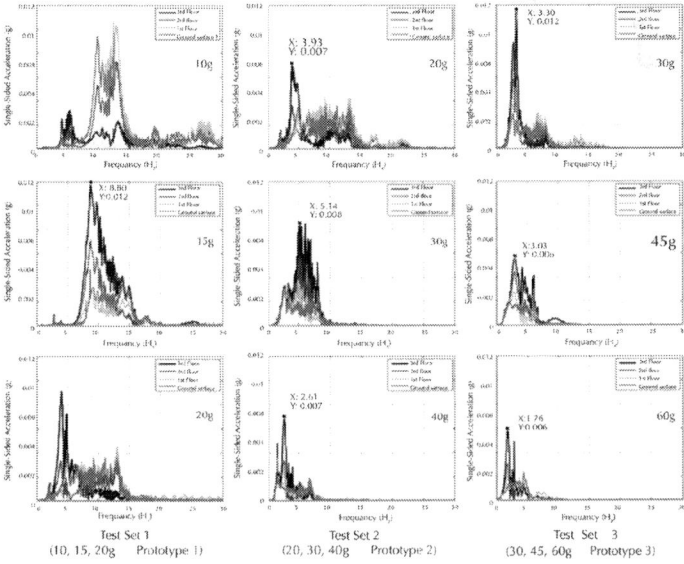

**Figure 6**: Frequency domain signals obtained from each test set for longitudinal excitation in the prototype scale.

To determine if these predominant frequencies could align with the natural frequencies of a real 3-storey stone structure, theoretical solutions for rocking and sliding motions of a single block were used for comparison. They are rocking [2] and [3] and sliding [4] and [5] motion solutions in the soil–foundation system. In this theoretical approach, it is assumed that the stone structure is composed of units with the same shape as that of a single block and there is no damping effect.

Solutions for the natural and resonant frequency of rocking vibration (Fig. 7(a)) for a rigid rectangular foundation are shown in Eqs. (1) and (2). Solutions for sliding vibration (Fig. 7(b)) are shown in Eqs. (3) and (4). Theoretically, the natural frequency of a 1/N model is N times the natural frequency of a prototype if scaling factors for mass (m) and length ($r_0$, D, W, H) are applied.

$$fn, prototype = \frac{1}{2\pi}\sqrt{\frac{k_\theta}{I_0}} = \frac{1}{2\pi}\sqrt{\frac{(8Gr_{0.p}^{\ 3}/3(1-\mu))}{m_p\left((r_{0.p}^{\ 2}/4)+H_p^{\ 2}/3\right)}}[2] \qquad (1)$$

$$fn, prototype = \frac{1}{2\pi}\sqrt{\frac{k_\theta}{I_0}} = \frac{1}{2\pi}\sqrt{\frac{(G/1-\mu)F_\theta D_p W_p^2}{m_p((r_{0.9}^2/4)+(H_p^2/3))}}[3] \qquad (2)$$

where\

$$r_{0.p} = \sqrt[4]{\frac{D_p W_p^3}{3\pi}}, F_\theta = \begin{cases} 0.6 \text{ at } W/D=2 \\ 0.44 \text{ at } W/D=0.5 \end{cases}$$

$$fn, prototype = \frac{1}{2\pi}\sqrt{\frac{k_x}{m_p}} = \frac{1}{2\pi}\sqrt{\frac{32(1-\mu)Gr_{0.p}}{(7-8\mu)m_p}}[4] \qquad (3)$$

$$fn, prototype = \frac{1}{2\pi} \sqrt{\frac{k_x}{m_p}} = \frac{1}{2\pi} \sqrt{\frac{2(1-\mu)GF_x \sqrt{D_p W_p}}{m_p}} \quad [5] \qquad (4)$$

where

$$r_{0.p} = \sqrt{\frac{D_p W_p}{\pi}} , F_x = \begin{cases} 0.95 \text{ at } W/D=2 \\ 1.1 \text{ at } W/D=0.5 \end{cases}$$

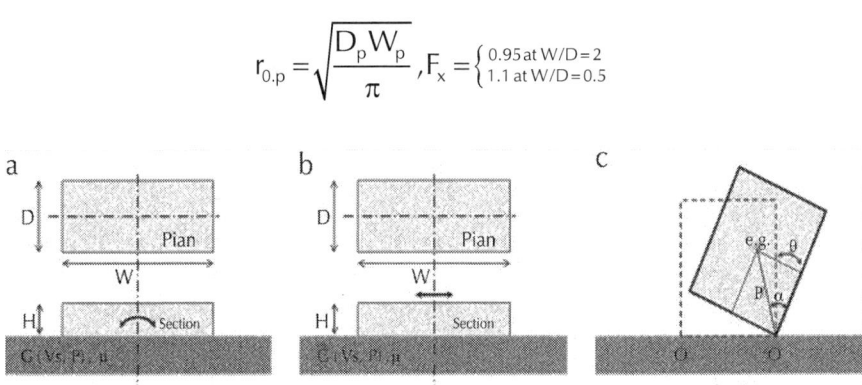

**Figure 7**: Theoretical approaches for rocking and sliding motion of single block: (a) rocking motion in the soil–foundation system; (b) sliding motion in the soil–foundation system; (c) Housner's rocking motion of single block structure.

The natural frequency for the rocking motion of a single block structure, according to Housner [7], is described by Fig. 7(c) and Eq. (5). Here, if a scaling law for acceleration of gravity (g) and a radius of rotation (R) is used, the natural frequency in a 1/N model becomes N times the natural frequency of the prototype, as in Eq. (6). In the case of multi-layered block structures, a certain scaling law should exist for various dynamic behaviours, including rocking, and this should be examined on an experimental basis. Housner's rocking motion assumes that relatively slender rigid blocks can oscillate about their centre of rotation (O), as shown in Fig. 7(c). Therefore, the theoretical solutions for the rocking and sliding motions of a single block in a soil–foundation system are more useful for comparison with experimental observation.

$$fn, prototype = \frac{\sqrt{m_p g_p R_p / I}_{0,p}}{4\cosh^{-1}(1/(1-\theta/\alpha))} = \frac{\sqrt{3g_p / 4R_p}}{4\cosh^{-1}(1/(1-\theta/\alpha))}[7] \qquad (5)$$

$$fn, model = \frac{\sqrt{3g_m / 4R_m}}{4\cosh^{-1}(1/(1-\theta/\alpha))} = \frac{\sqrt{3(Ng_p)/4(R_p/N)}}{4\cosh^{-1}(1/(1-\theta/\alpha))} = Nfn, prototype \qquad (6)$$

Fig. 8 shows the comparison between the observed amplified frequency (i.e. predominant frequency) range in all test sets and the theoretical natural frequency for rocking and sliding motions of a single block. The upper figures show transversal excitation and the lower ones show longitudinal excitation. The predominant frequency and amplified range from the test results and the natural frequency using theoretical equations are tabulated as well. It was assumed that the shear wave velocity of the ground, which was considered to be compacted weathered soil, was 200 m/s; mass density, 1800 kg/m³; and Poisson's ratio, 0.33.

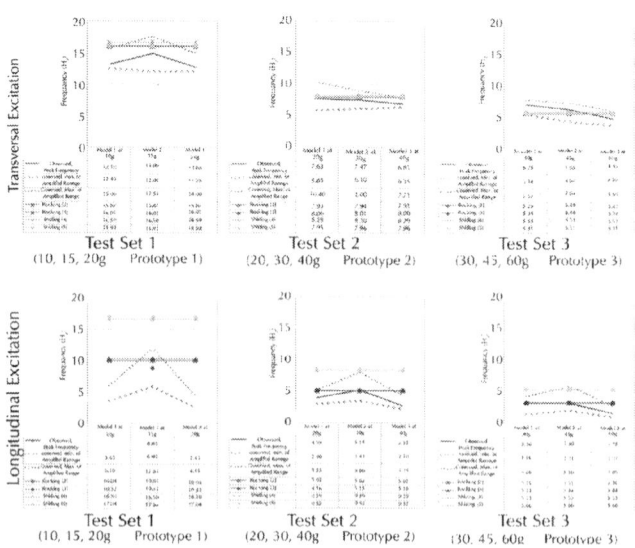

**Figure 8**: Comparison between observed predominant frequency range and theoretical natural frequency for rocking and sliding motion of single block in prototype scale.

In the case of transversal excitation, the natural frequency ranges of rocking and sliding had theoretically similar values. The predominant frequency range of the test results, after application of the scaling factors, was also within a similar range. On the other hand, for longitudinal excitation, the natural frequencies of rocking and sliding had different ranges. The predominant frequency range of the test results was closer to the theoretical rocking motion range.

The earthquake simulator that was used in the tests has an excitation frequency range of 40–300 Hz, at the model scale. Therefore, when performing the 'modelling of models' procedure, if the same seismic wave is excited for each g-level, the transformed frequency domain will not coincide exactly at the full-size prototype scale. For example, the shaking energy exists in the domain of 4–30 Hz at the 10g level and in the domain of 2–15 Hz at the 20g level. Fig. 9 shows the frequency ranges that can be compared for each test set. 3 models were set up in 3 test sets because it is difficult to predict the resonant frequency of masonry structures in advance and the excitation frequency ranges of the earthquake simulator are predetermined.

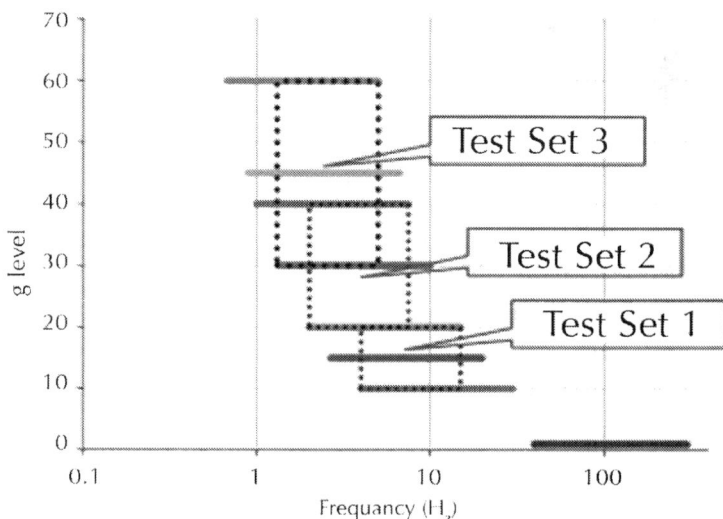

**Figure 9**: Frequency ranges that can be compared for each test set.

For determination of the model size in a dynamic centrifuge test, the predicted natural frequency range of the target structure, excitation frequency ranges of the earthquake simulator, maximum operational acceleration of the centrifuge, maximum shaking acceleration of the earthquake simulator, mass of the model, and the size of the container must be systematically considered.

Through these test results and theoretical comparison, it was revealed that acceleration amplification characteristics and that predominant frequency ranges of the 3 storey block structure coincided between theoretical predictions and observed results. Additionally, this proves that the generalised scaling law can be applied to the dynamic behaviours of masonry structures.

# CONCLUSIONS

A 3-storey stone block structure was built at 3 different scales and the 'modelling of models' procedure was carried out to evaluate for seismic behaviour of the scaled structure using an earthquake simulator within a centrifuge. The dynamic centrifuge tests were conducted at different g-levels so that the behaviour of one full-size imaginary prototype could be simulated and the effect of scaling compared. The test revealed that similar behaviours appeared within a range of error when the results were transformed to a full-size prototype scale.

This research proves the possibility of predicting the seismic behaviour of an actual structure using a small-scale model in a centrifugal acceleration field when the proper scaling laws are applied. Therefore, it has been confirmed that the seismic behaviour and risk of future damage to historic stone monuments can be assessed with dynamic centrifuge tests of scaled models.

# ACKNOWLEDGEMENTS

This research was supported by a Basic Science Research Program through the National Research Foundation of Korea (NRF) funded by the Ministry of Education, Science and Technology (grant number: 2009-0080575). The authors gratefully acknowledge the KREONET service provided by Korea Institute of Science and Technology Information.

# REFERENCES

1.  R.N. Taylor, Geotechnical centrifuge technology, Blackie Academic and Professional, London (1995)

2.  Borowicka, H., Uber Ausmittig Belastete Starre Platten auf Elastischisotropem Untergrund, Ingenieur-Archiv, Berlin, 1 1943 p. 1–8.

3.  Gorbunov-Possadov MI, Serebrajanyi, RV. Design of structures upon elastic foundations, In: Proceedings of the 5th international conference on soil mechanics and foundation engineering, Paris, vol.1: 1961. p. 643–8.

4.  G.N. Bycroft, Forced vibrations of a rigid circular plate on a semi-infinite elastic space and on an elastic stratum, Philosophical Transactions of the Royal Society of London Series A, 248 (1956), pp. 327–368

5.  D.D. Barkan, Dynamic bases and foundations, McGraw-Hill Book Company, New York (1962)]

6.  B.M. Das, G.V. Ramana, Principles of soil dynamics, (2nd ed.) Cengage Learning (2010)

7.  G. Housner, The behavior of inverted pendulum structures during earthquakes, Bulletin of the Seismological Society of America, 53 (2) (1963), pp. 403–417

8.  Y. Ishiyama, Motions of rigid bodies and criteria for overturning by earthquake excitations, Earthquake Engineering and Structural Dynamics, 10 (5) (1982), pp. 635–650

9.  P.D. Spanos, A.S. Koh, Rocking of rigid blocks due to harmonic shaking, Journal of Engineering Mechanics, 110 (11) (1984), pp. 1627–1642

10. W. Tso, C. Wong, Steady state rocking response of rigid blocks. Part 1: analysis, Earthquake Engineering and Structural Dynamics, 18 (1) (1989), pp. 89–106

11. I Psycharis, Dynamic behaviour of rocking two-block assemblies, Earthquake Engineering and Structural Dynamics, 19 (4) (1990), pp. 555–575

12. N. Makris, J. Zhang, Rocking response of anchored blocks under pulse-type motions, Journal of Engineering Mechanics, 127 (5) (2001), pp. 484–493

13. P.D. Spanos, P.C. Roussis, N.P.A. Politis, Dynamic analysis of stacked rigid blocks, Soil Dynamics and Earthquake Engineering, 21 (7) (2001), pp. 559–578

14. N. Makris, D. Konstantinidis, The rocking spectrum and the limitations of practical design methodologies, Earthquake Engineering and Structural Dynamics, 32 (2) (2003), pp. 265–289

15. I.N. Psycharis, D.Y. Papastamatiou, A. Alexandris, Parametric investigation of the stability of classical columns under harmonic and earthquake excitations, Earthquake Engineering and Structural Dynamics, 29 (8) (2000), pp. 1093–1109

16. I.N. Psycharis, J.V. Lemos, D.Y. Papastamatiou, C. Zambas, C. Papantonopoulos, Numerical study of the seismic behaviour of a part of the Parthenon Pronaos, Earthquake Engineering and Structural Dynamics, 32 (13) (2003), pp. 2063–2084

17. D. Konstantinidis, N. Makris, Seismic response analysis of multidrum classical columns, Earthquake Engineering and Structural Dynamics, 34 (10) (2005), pp. 1243–1270

18. J.K. Kim, H. Ryu, Seismic test of a full-scale model of a five-storey stone pagoda, Earthquake Engineering and Structural Dynamics, 32 (5) (2003), pp. 731–750

19. D'Ayala D, Shi Y, Stammers, C.. Dynamic multi-body behaviour of historic masonry buildings models, structural analysis of historic construction: preserving safety and significance, two volume set. In: Proceedings of the VI international conference on structural analysis of historic construction, SAHC08, 2–4 July, Bath, United Kingdom, 2008. p. 489–6.

20. F. Peña, P.B. Lourenço, A. Campos-Costa, Experimental dynamic behavior of free-standing multi-block structures under seismic loadings, Journal of Earthquake Engineering, 12 (6) (2008), pp. 953–979

21. D. Konstantinidis, N. Makris, Experimental and analytical studies on the response of 1/4-scale models of freestanding laboratory equipment subjected to strong earthquake shaking, Bulletin of Earthquake Engineering, 8 (6) (2010), pp. 1457–1477

22. A.N. Kounadis, G.J. Papadopoulos, D.M. Cotsovos, Overturning instability of a two-rigid block system under ground excitation, ZAMM—Journal of Applied Mathematics and Mechanics/

Zeitschrift fur Angewandte Mathematik und Mechanik, 92 (7) (2012), pp. 536–557

23. D. D'Ayala, A. Ansal, Non-linear push over assessment of heritage buildings in Istanbul to define seismic risk, Bulletin of Earthquake Engineering, 10 (1) (2012), pp. 285–306

24. S. Iai, T. Tobita, T. Nakahara, Generalized scaling relations for dynamic centrifuge tests, Geotechnique, 55 (5) (2005), pp. 355–362

25. D.S. Kim, N.R. Kim, Y.W. Choo, G.C. Cho, A newly developed state-of-the-art geotechnical centrifuge in Korea, KSCE Journal of Civil Engineering, 17 (1) (2013), pp. 77–84

26. D.S. Kim, S.H. Lee, Y.W. Choo, J. Perdriat, Self-balanced earthquake simulator on centrifuge and dynamic performance verification, KSCE Journal of Civil Engineering, 17 (4) (2013), pp. 651–661

27. Stokoe KH, Hwang SK, Roesset JM, Sun WS. Laboratory measurements of small-strain material damping of soil using the free–free resonant column, In: Proceedings of the earthquake resistant construction and design conference, Berlin, Germany, , Rotterdam: Balkema; 1994. p. 195–202.

# Chapter 4

# Formation of Tabular Plutons – Results and Implications of Centrifuge Modelling

Carlo DIETL[1] and Hemin KOYI[2]

[1]Institut für Geowissenschaften, Goethe-Universität, Altenhöferallee 1, D-60438 Frankfurt amMain, Germany

[2]Hans Ramberg Tectonic Laboratory, Department of Earth Sciences, Uppsala University, Villavägen 16, S-75236 Uppsala, Sweden

## INTRODUCTION

Field geological as well as geophysical investigations on numerous large plutons (encompassing areas in the range of several hundred up to more than thousand km²) during the last 20 years have revealed that many of them are of tabular, either laccolithic (convex roof and straight floor)

or lopolithic (straight roof and concave floor) or phacolithic (biconvex) shape. Despite their differences in emplacement level and tectonic setting, the development of the lensoidal shapes of all these plutons can be described as a three-stage process (e.g. Corry 1988; Cruden 1998). In a first stage a vertical dyke is established, where magma is transported upward until it reaches a horizontal unconformity along which the magma then – in the second stage – spreads laterally to form a sill. In the third stage additional magma ascends through the feeder dyke to be emplaced into the sill, which inflates due to the magma overpressure by doming of the roof and/or depression of the floor of the growing pluton. These three stages are incremental processes and intimately entangled with each other, in particular in the case of composite plutons which consist of several magma batches. Magma ascent generated by dyking as well as roof uplift and floor depression of the developing pluton take advantage of shear zones and faults; either as magma pathways (magma ascent) or as planes of movement along which individual blocks of the roof and floor of the evolving pluton are shifted upward or downward (magma emplacement). The lateral spreading of magma prior to magma chamber inflation benefits from horizontal anisotropy planes in the host rock, such as bedding and foliation planes, fractures or shear zones. According to Cruden (1998) the roof uplift plays a major role as a space-making process only for laccoliths emplaced in the uppermost crust at around 3 km depth, while floor depression seems to be responsible for most of the – then lopolithic – tabular plutons being emplaced deeper within the Earth's crust below c. 4–5 km depth.

Based on dimensional data from several hundred plutons McCaffrey and Petford (1997) and Petford and Clemens (2000) showed that the proportions of intrusive bodies (length vs. thickness) can be described by the following power law:

$$T = c \times L^a \qquad\qquad (1)$$

with pluton thickness $T$, its length $L$, a constant $c$ and the power-law exponent $a$. McCaffrey and Petford (1997) based their calculations on field data from 135 laccoliths and 21 plutonic intrusions. According to their data base power law (1.1) for laccoliths only can be formulated as:

$$T=0.12\times L^{0.88} \tag{1.2}$$

while power law (1.1) applied to plutons is

$$T=0.29\times L^{0.80} \tag{1.3}$$

Cruden (1998) ended up with a value of 0.29 for constant $c$ when calculating the power law relationship between the length and thickness of tabular plutons. Petford and Clemens (2000) found – based on the dimensional data from more than 100 plutons and batholiths – that the power law exponent $a$ in formula (1.1) equals 0.6 ± 0.1. According to Petford et al. (2000), tabular plutons, which are established mainly by downward directed material transfer (e.g. floor depression) can be distinguished from laccoliths from the power-law regression lines they follow in a loglength vs. logthickness plot. Laccoliths plot along a power-law regression line with $a = 0.88 ± 0.1$, while plutons appear in this diagram along regression line with $a = 0.60 ± 0.1$ (Fig. 1).

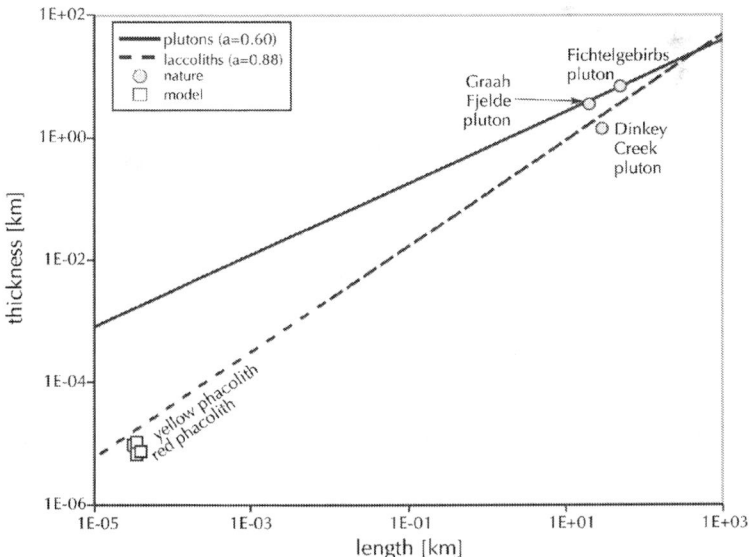

**Figure 1:** Length vs. thickness plot for plutons and laccoliths, showing their differing dimensional ratios as expressed by the two regression lines (Pet-

ford et al. 2000). Grey dots indicate length-thickness ratios of the three real plutons discussed in text, open squares show the dimensions of the model phacoliths.

The differences in numbers for *a* and *c* reflect the differing data sets the individual power law formulas are based on and the quality of these data. Moreover, for the formulation of the individual equations only two – thickness and length – of the three available dimensional data (thickness, length and width) are taken into account. However, in particular the thickness data for the individual intrusions are not well constrained, because until today no major intrusive body with exposed roof *and* floor was observed. In particular descriptions of outcropping intrusion floors lack almost entirely in literature. One exception is the Bergell pluton in the Central Alps, the floor of which is exposed, together with its root zone (Rosenberg et al. 1995). Consequently, information about intrusion floors stem mainly from indirect, geophysical methods, such as gravimetry and seismic investigations. Nevertheless, all these six power laws for laccoliths and plutons have one thing in common: the power law exponent *a* is always < 1, indicating the tabular shape of the granitoid bodies. However, the power law behaviour of these tabular intrusives does not only describe their shape, but also their growth: they start as thin, horizontal sills, which only after some lateral growth expand also vertically, giving them their final tabular shape.

Dinkey Creek pluton (Sierra Nevada, California)

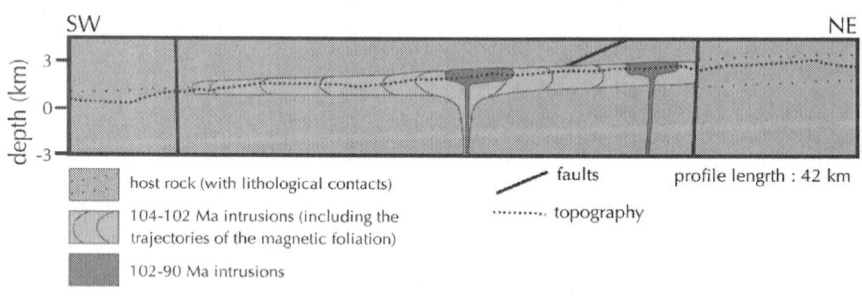

**Figure 2:** Cross section through the Cretaceous Dinkey Creek pluton (Sierra Nevada, redrawn from Cruden et al. 1999).

# NATURAL EXAMPLES FOR TABULAR PLUTONS

Typical tabular plutons include the upper-crustal, synorogenic Cretaceous Dinkey Creek pluton in the Sierra Nevada (California; Cruden et al. 1999; Hammarstrom and Zen 1986; Fig. 2), the mid-crustal, late-orogenic Carboniferous Fichtelgebirge pluton (Bohemian Massif; Hecht and Vigneresse 1999; O'Brien 2000; Fig. 3), and the upper-crustal, late- to post-orogenic Proterozoic Graah Fjelde Rapakivi Granite (Greenland – Grocott et. al. 1999; Garde et al. 2002; Fig. 4). Grocott et al. (1999), Hecht and Vigneresse (1999) and Cruden et al. (1999) suggested floor depression by the inflowing magma due to its overpressure as the main emplacement mechanism for the Graah Fjelde, Fichtelgebirge and Dinkey Creek plutons, respectively. All three tabular intrusions named here follow the power-law for pluton proportions developed by McCaffrey and Petford (1997) and Petford and Clemens (2000). Plotting the thickness-length ratios of the Graah Fjelde, Fichtelgebirge and Dinkey Creek plutons, respectively (calculated based on the cross sections shown in Figs 2–4; see also Tab. 1) reveals that the Graah Fjelde and Fichtelgebirge plutons occur close to the "pluton" regression line and the Dinkey Creek pluton slightly below the "laccolith" regression line (Fig. 1). According to the interpretation of Petford et al. (2000) the Graah Fjelde and Fichtelgebirge plutons should be regarded as lopoliths emplaced by floor depression; and the Dinkey Creek pluton should then be interpreted as a laccolith emplaced by roof-uplift, although all three intrusive bodies plot quite close to the intersection of both the regression lines.

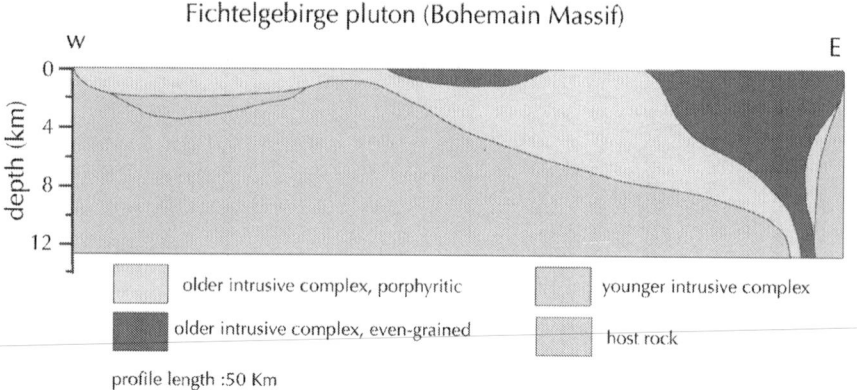

**Figure3:** Cross section through the Variscan Fichtelgebirge pluton (Bohemian Massif, redrawn from Hecht and Vigneresse 1999). The older and younger intrusive complexes are genetically not related to each other (J.L. Vigneresse, pers. comm.

## Graah Fjelde rapakivi granite complex, (E. Greenland)

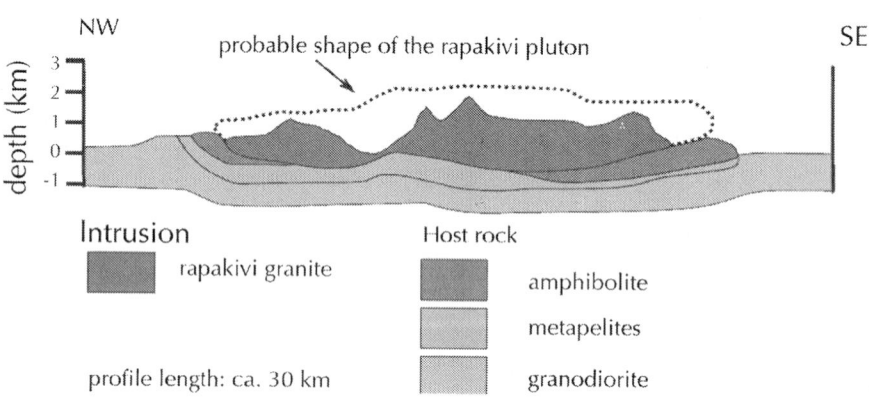

**Figure 4:** Cross section through the Precambrian Graah Fjelde Rapakivi Granite (Greenland, redrawn from Garde et al. 2002).

**Table 1:** Dimensions of the three presented plutons as taken from Figs2–4, as well as of the two PDMS intrusions in the centrifuge model

| Natural examples | Thickness [km] | Length [km] |
|---|---|---|
| | 1.38 | |
| Dinkey Creek pluton | 1.38 | 29.22 |
| Fichtelgebirgs pluton | 6.82 | 50.00 |
| Graah Fjelde pluton | 3.47 | 20.06 |
| PDMS intrusions | Thickness [cm] | Length [cm] |
| Yellow phacolith, section 1 | 0.96 | 3.24 |
| Yellow phacolith, section 2 | 1.08 | 3.50 |
| Red phacolith, section 1 | 0.68 | 3.48 |
| Red phacolith, section 3 | 0.75 | 4.001 |

# THE EXPERIMENT

## Motivation of the Experiment

The three examples presented briefly above are only a small selection of tabular plutons, which are common features of the continental crust. They develop independently of their temporal, tectonic or rheological context, usually according to the same ascent and emplacement style, i.e. dyking, followed by sill formation and magma chamber inflation. According to McCaffrey and Petford (1997), their growth during emplacement is ruled by the power law (1.1). Consequently, also the dimensions of a growing pluton follow the power law (1.1): it lengthens more rapidly than it thickens.

Analogue modeling has proven to be an effective tool in studying geologic structures and processes (Koyi 1997). Here we present the results of an analogue centrifuge experiment, to test the viability of the proposed emplacement mechanisms for tabular plutons, in particular:

- how roof uplift and floor depression are linked to each other;
- whether floor depression is more typical of deep-seated and roof uplift more common of shallowly intruding tabular plutons;
- whether floor depression might hinder the filling and inflation of the magma chamber due to chocking of the magma source

(provided the magma source is a horizontal layer of partially molten rock).

Moreover, we test the validity of the assumptions made by Petford et al. (2000) concerning the discrimination between (lopolithic) plutons and laccoliths based on power-law regression lines in length-thickness diagrams for intrusive bodies.

# Setup of the experiment

The overall shape of the model was cylindrical with a diameter of 10 cm and 6.75 cm thick. The scaling factor for distances is 10-5 (lengthmodel / lengthnature), i.e. 1 cm in the model corresponds to 1 km in nature (Tab. 2). As overburden material a semi-brittle plasticine (density $\rho$ = 1.71 g/cm$^3$, viscosity $\mu$ = 4.2 × 107 Pa·s at a strain rate of $10^{-3}$ s$^{-1}$) was used, representing upper to middle crustal siliciclastic rocks ($\rho \approx 2.6$ g/cm3 and $\mu \approx 1021$ Pa·s). Polydimethylsiloxane (PDMS; $\rho$ = 0.964 g/cm$^3$, $\mu$ = 4×104 Pa·s; Koyi 1991) played the role of the buoyant material, corresponding to a partially molten granitic magma ($\rho \approx 1.5$ g/cm3, $\mu \approx 1018$ Pa·s). The assumed crustal viscosity of 1021 Pa·s is quite low, but falls within the range of presumed viscosity values for crustal rocks of 1020 to 1023 Pa·s (Koyi et al. 1999). These estimated viscosity values for the upper and middle crust are based

**Table 2:** Dimensions and physical parameters of both the model and its natural counterpart as well as their ratios (t = thickness, $\rho$ = density, $\mu$ = viscosity)

| Scaling | Model | Nature | Ratio (model/nature) |
|---|---|---|---|
| $t_{overburden}$ (red PDMS source layer) | 3.2 cm | 3.2 km | $10^{-5}$ |
| $t_{overburden}$ (yellow PDMS source layer) | 6.05 cm | 6.05 km | $10^{-5}$ |
| $t_{buoyant}$ | 0.5 cm | 500 m | $10^{-5}$ |
| $\rho_{overburden}$ | 1.71 g/cm$^3$ | 2.63 g/cm$^3$ | $6.5 \times 10^{-1}$ |
| $\rho_{buoyant}$ | 0964 g/cm$^3$ | 1.48 g/cm$^3$ | $6.5 \times 10^{-1}$ |
| $\mu_{overburden}$ | $4 \times 10^7$ Pa•s | $10^{21}$ Pa•s | $4.2 \times 10^{-14}$ |

| $\mu_{buoyant}$ | $4 \times 10^4$ Pa•s | $10^{18}$ Pa•s | $4.2 \times 10^{-14}$ |
|---|---|---|---|
| g | $6.87 \times 10^3$ m/$s^2$ | 9.81 m/$s^2$ | $7 \times 10^2$ |

on experimental determinations of effective viscosities for dry and wet quartzites at a temperature between 300 °C and 700 °C and at a geologically relevant strain rate of 10-14 s$^{-1}$ (Weyermars 199  and references therein), which gave effective viscosities between 1021 Pa·s (wet and hot quartzite) and 1023 Pa·s (dry and cold quartzite). On the other hand, the assumed viscosity of the partially molten granitic magma – calculated on the basis of the PDMS viscosity – is much too high compared to experimentally determined viscosity values of granitic magmas, which range between 103 and 1012 Pa·s (Dingwell et al. 2000). However, applying proper scaling and using consequently fluids with very low viscosities, such as water ($\mu \approx$ 10-3 Pa·s) or olive oil ($\mu \approx$ 10-1 Pa·s) as magma analogues would lead to ascent and emplacement structures which are of unrealistic shape for granitic intrusive bodies (Ramberg 1981). As such, the presented model is only partially scaled to nature. However, the model is a first approximation to nature and focuses on the formation of tabular intrusions and the effects of floor depression and roof uplift on their ascent and emplacement in general.

Model stratigraphy included two buoyant layers of PDMS (5 mm thick each and with a volume of c. 40 cm3 each) at the base and in the middle (3 cm high in the model stratigraphy) of the model. At the upper boundary of each buoyant layer a perturbation (1 cm × 1 cm × 0.7 cm) was introduced to trigger the formation of two intrusions. Of course, it is impossible to model with these two single intrusions exactly the incremental growth of a tabular pluton due to the continuous influx of magma. However, the presented model serves as a good proxy for the role roof uplift and floor depression play during the ascent into and the construction of a tabular granitoid body due to the rise of a single magma batch. The perturbation of the lower buoyant layer was placed in the centre of the model (5 cm from its rim), while the perturbation of the middle buoyant layer was shifted from the model's centre by 2.3 cm (Fig. 5). Both the perturbations were intended to act as pathways for the ascending buoyant material during centrifuging, in correspondence to the feeder dykes of tabular plutons. The horizontal offset between both the perturbations was designed to prevent the developing intrusive bodies to interact too much with each other. The overburden of both the buoyant layers was stratified consisting of five (above the lower buoyant layer) and eight (above the upper buoyant layer) differently coloured plasticine layers of 0.25 to 0.7 cm thick (Fig. 5).

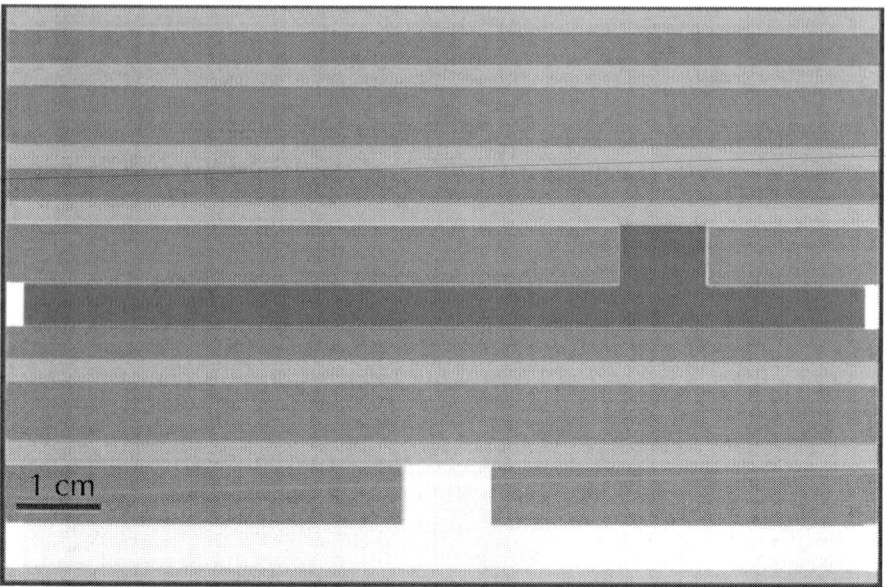

1 cm

**Figure 5:** Setup of the experiment. The yellow and red layers represent the buoyant PDMS, the light and dark green as well brown layers are the plasticine overburden.

## Run of the experiment

The model was centrifuged in the small centrifuge of the Hans Ramberg Tectonic Laboratory (Uppsala University, Sweden) for c. 5 minutes at 700 G when a small bulge formed on top of the model directly above the perturbation of the middle buoyant layer. After photographing the top view with the small bulge on the model's surface (Fig. 6), the model was centrifuged for 20 more minutes at 700 G in order to see, whether in the current configuration the bulge on top of the model would increase or a second bulge would form. When no major changes happened after 25 minutes of centrifuging, the uppermost three plasticine layers – in total 0.9 cm of the overburden – were removed to thin the overburden and assist the rise of the buoyant material. The model was then centrifuged for further 5 minutes at 700 G. No major changes in the top view of the model occurred and the model was then sectioned in order to study it in detail.

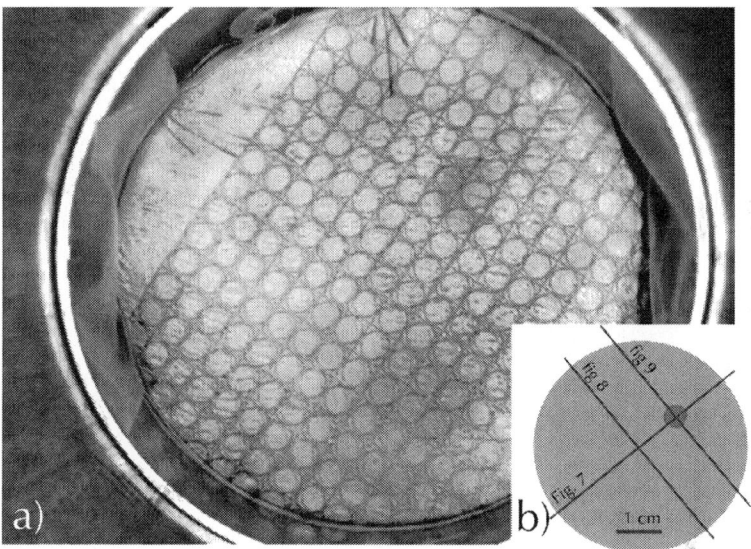

**Figure 6:** (a) Top view of the model after 5 minutes of centrifuging at 700 G. The slight bulge northeast (with north at the top of the image) of the centre marks the location of the upper phacolith. The diameter of the model is 10 cm. **Inset b** – sketch of the photograph in Fig. 6a with the location of the red phacolith indicated by the red circular area above and to the right of the centre of the model (model surface in green); the lines indicate the positions of the sections presented in Figs 7–9.

## Results of the experiment

A first vertical section in the middle of the model showed two tabular intrusive bodies that had formed directly above the perturbations along the interface between the buoyant layers and the overburden layers immediately overlying them (Fig. 7). Both the intrusions are characterized by a domed-up roof and a depressed floor, which is typical of phacoliths. The subsided floors of both PDMS bodies cut off the influx of the buoyant material from the source layers. The overburden units above the lower phacolith bulge into the middle buoyant layer, which accommodates the deformation by thinning (Fig. 7). The plasticine layers initially located directly above the buoyant layers, partly detach from the overlying overburden layers and subside in the immediate vicinity of the perturbations (Fig. 7). Moreover, the plasticine layer immediately beneath the perturbation of the upper buoyant layer is detached from its pad to form a small bulge (Fig. 7).

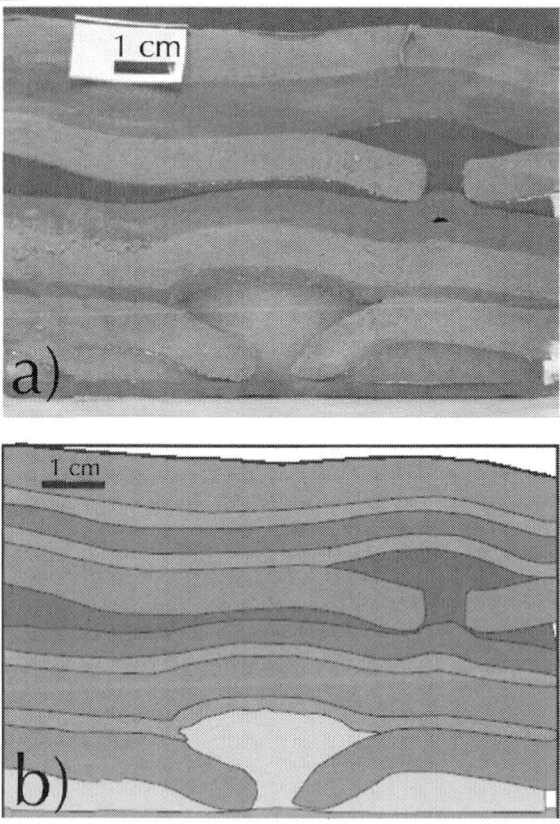

**Figure 7:** a – Section 1 through the model after 30 minutes of centrifuging at 700 G cutting both phacoliths: the red one in the upper part of the model and the yellow one in its lower part. b – line drawing of the section shown in Fig. 7a.

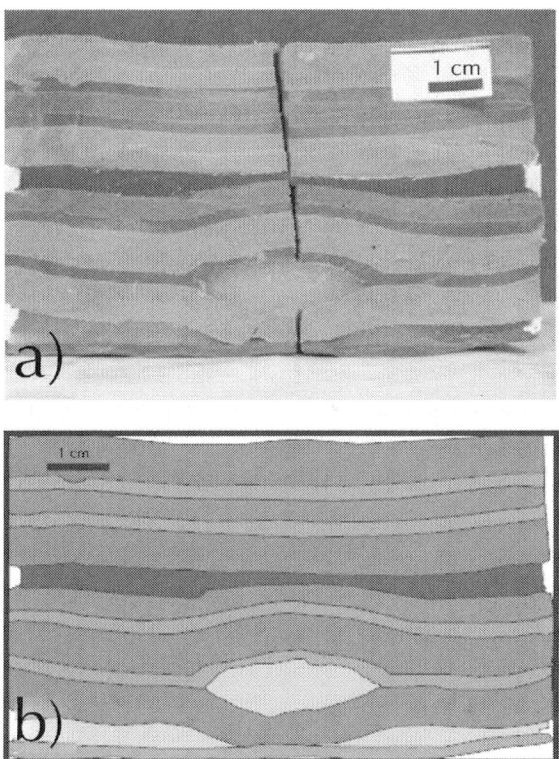

**Figure. 8**a – Section 2 – perpendicular to section 1 – through the model after 30 minutes of centrifuging at 700 G proofs the lensoidal shape of the yellow intrusive PDMS body. b – line drawing of the section shown in Fig. 8a.

Two additional profiles (Figs 8 and 9), both perpendicular to the first one and carving each through one of the two buoyant intrusions, verified the phacolith nature of both bodies and helped to measure their dimensions and, in addition, calculate their volumes. The lower phacolith is between 3.24 and 3.5 cm in diameter, depending on which section is considered. Its maximum thickness lies between 0.96 and 1.08 cm. The upper phacolith is 3.48 to 4.00 cm long and 0.68 to 0.75 cm thick (Tab. 1). The volume of the tabular buoyant bodies was calculated according to the following equation (Bronstein and Semendjajew 1987):

$$V = 2 \times (1/6)\pi \times h(3a^2 + h^2) \qquad\qquad (2)$$

where $h$ is half the height and $a$ is half the diameter of the respective phacoliths.

Accordingly, the volume of the lower intrusion is between 4.06 and 5.34 cm$^3$ and the volume of the middle intrusion ranges between 3.29 and 4.77 cm$^3$.

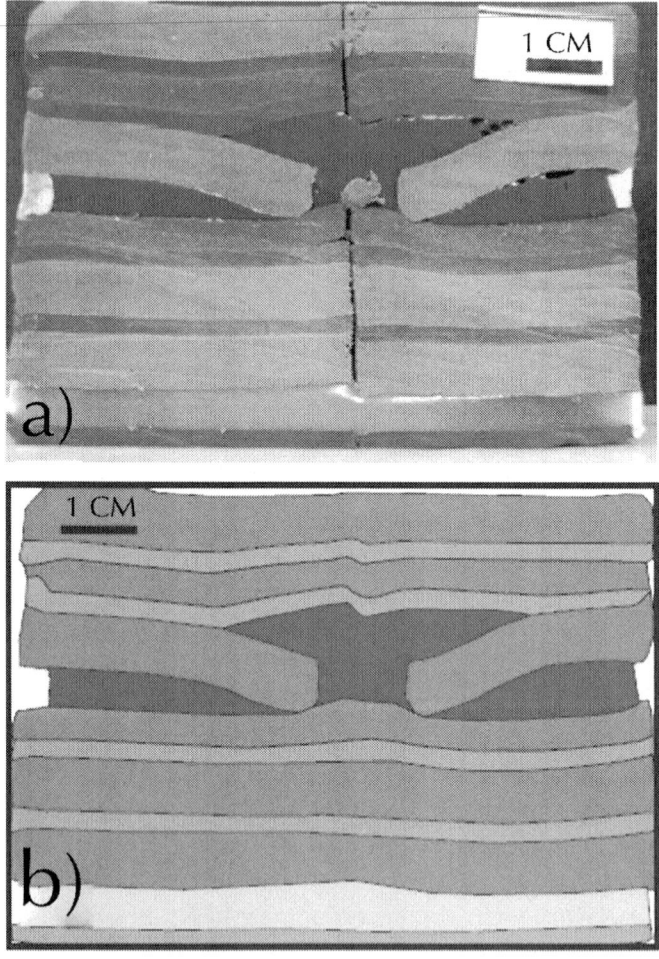

**Figure 9a:** Section 3 – perpendicular to section 1 and parallel to section 2 – through the model after 30 minutes of centrifuging at 700 G proofs the

lensoidal shape also of the red intrusive PDMS body. **b** – line drawing of the section shown in Fig. 9a.

# FORMATION OF BOTH THE PHACOLITHS

As planned, the perturbation in the buoyant layers acted as pathways (or feeder dykes) for the buoyant material during centrifuging of the model. However, the buoyant material did not manage to rise through the plasticine overburden and instead spread laterally as sills along the interfaces between the two plasticine layers directly above the buoyant layers. With further centrifuging, additional material was transported buoyantly from the PDMS layer through the feeder dykes into the sills. The influx of new material led to inflation of the sills by vertical displacement of their roofs and floors. Thereby the plasticine above the developing phacoliths was deformed plastically within a distance of 1.5 to 2 cm, while the floors of the inflated intrusions were depressed by c. 0.5 cm. Additionally, the thin plasticine layer beneath the upper buoyant layer was sucked up into the conduit of the buoyant body by the upward moving buoyant material. On the other hand, subsidence of the bottom of the inflating sills choked the flow of further buoyant material into the feeder dykes and inhibited the further growth of the two phacoliths. Inflation of the lensoidal intrusives terminated after c. 5 min centrifuging when no increase of the bulge on the surface of the model was observed. Until then about 4 to 5 cm³ of the buoyant material had been emplaced into the tabular intrusive bodies, i.e. the buoyant reservoirs had been depleted by 10 to 12.5 vol. %.

It can be excluded that the formation of both intrusive PDMS bodies through roof uplift and floor depression were influenced by boundary effects such as the relatively small volume of the sample box, because both lensoidal intrusions and their structural aureoles look very much alike irrespective from their relative position within the model and the sample box. Moreover, the total volume of the intrusions is only c. 10 cm³, i.e. less than 2 % of the entire model volume (c. 530 cm³).

# IMPLICATIONS OF THE EXPERIMENT FOR THE EMPLACEMENT OF TABULAR PLUTONS

Scaling the model to nature the lensoidal intrusive buoyant bodies would represent 1 km thick and 3 to 4 km long granitoid phacoliths (Tabs 1 and 2) that were emplaced at 3.2 km and 6.05 km depths, respectively, i.e. at different levels within the upper continental crust.

Both tabular intrusions formed – similar to what is known from their natural counterparts – along anisotropies within the model, i.e. along the interfaces between individual overburden layers. Corresponding anisotropies in nature are horizontal lithological contacts, foliation and fracture planes or shear zones.

The phacolithic shape of the intrusions suggests that space for their emplacement was made by both roof uplift and floor depression, independent of their intrusion depth. This observation contradicts the hypothesis of Cruden (1998) who suggested roof uplift to be only an important space-making process for plutons in the uppermost crust, while floor depression is responsible for the emplacement of all other, deeper-seated tabular plutons. We have to point out that the reason behind achieving roof uplift by the lower intrusion is because of the presence of another weak buoyant layer in the middle of the model stratigraphy. This weak layer enabled and accommodated the roof uplift at deeper level. In other words, in our model, similar to a shallow intrusive body, even the deeper intrusive body uplifted its roof. This was possible because the roof uplift of the deeper intrusion was accommodated by flow of the upper buoyant (viscous) layer. In practice, the mechanical thickness of the overburden units above the lower buoyant layer is relatively thin. The presence of the upper buoyant weak layer enables and accom- modates roof uplift by the deeper intrusion. Absence of this shallow buoyant weak horizon would have resulted in a thicker overburden unit, which might not have been uplifted by a deep intrusion. Presence of layered intrusions may assist roof uplift even at deeper levels. The same is in principle true for floor depression which was facilitated by the presence of the two relatively low-viscous PDMS source layers beneath the two developing lensoidal PDMS intrusions.

Plotting the length and thickness data of the PDMS laccolith in the loglength vs. logthickness diagram of Petford et al. (2000) with the regression lines for laccolith and tabular plutons (Fig. 1) the data pairs of both sections of the model appear close to the regression line for laccoliths. The model intrusions are definitely not laccolithic in shape and were, consequently, not only emplaced by roof uplift – as typical of laccoliths (Figs 7–9 and Tab. 1). On the contrary, they are biconvex in shape and were intruded by a combination of roof uplift and floor depression – as it is typical of phacoliths. Obviously, more data from nature and models – in particular reliable thickness data for granitoid bodies – are necessary to improve the regression lines which could then describe better the shape of different types of intrusions.

Displacement of the top and bottom sides of both the experimental tabular intrusions took place plastically. No brittle deformation or concentration of deformation along discrete fractures, faults or shear zones was observed as it might be the case for most natural lacco-lopo- or phacoliths (e.g. Bussel 1976; Dehls et al. 1998; Zimmerman 2005; Hacker et al. 2007).

The most prominent feature within the emplacement scenario of the model is the choking of the influx of buoyant material into the feeder dykes of both the phacoliths due to the floor depression of the inflating intrusive bodies. The source layer of the buoyant material was segmented by the subsiding overburden units and further flow into the intrusive body was inhibited. As such, the scenario presented by Cruden (1998), where the depression of the floor of an inflating tabular pluton will force the magma within its source region towards the conduit(s) of the growing pluton, does not apply to our model. We argue therefore that Cruden's (1998) idea depends on the initial geometry of the feeding magma source; a horizontal magma source, possibly a migmatitic layer within the Earth's crust would face the fate as the PDMS source layers of our model, whereas a magma source with an irregular geometry would most likely behave as Cruden (1998) proposed. Nevertheless, the "strangulation" of the buoyant-material supply in the model did not prevent the formation of the intrusive bodies, but restricted their growth to a volume of 4 to 5 cm3 (reached within c. 5 minutes) and limited the depletion of the buoyant reservoir to c. 10 vol. %. Transferring these model observations to nature could mean that the size of tabular plutons and their growth rate are not only controlled by the volume of the reservoir and the rate of magma ascent and the magma's physical properties ($T$, $\mu$, $\rho$), but also by the shape of the magma source region and, moreover, by the effectiveness of vertical, downward-directed material transfer to create space for the build-up of the pluton. Time of the model phacoliths (tm) is scaled to nature (tn) as follows (Ramberg 1981):

$$t_n = (t_m \times l_r \times p_r \times g_r) / \mu_r \tag{3}$$

where $l_r$ is the length ratio, $_r$ is the density ratio, $g_r$ is the gravitational acceleration ratio and $\mu_r$ is the viscosity ratio between model and nature (for numbers see Tab. 2). Even if the model is dynamically not properly scaled, because of the too high viscosity of the buoyant material compared to its overburden, time scaling based on the available numbers (Tab. 2) gives a rough estimate of the time span for the formation of a natural phacolith before its growth is stopped due to choking of the influx of more magma by floor depression.

In the model, it took c. 5 minutes to form a lensoidal intrusion of 4 to 5 cm3 0.7 cm above the buoyant source layer. Accordingly, in nature a phacolith of 4 to 5 km³, being emplaced 700 m above its source layer, would have formed within c. 1 Ma. This time span would be sufficient to emplace such small tabular plutons. Assuming a lower and more realistic viscosity of the magma (109 to 103 Pa·s, e.g. Clemens and Petford 1999) than that used in the model (1018 Pa·s) more voluminous tabular intrusive bodies could be built in a greater distance from the magma source layer within the same or even a shorter time. Ascent rates of magma through dykes are in the range of 106 to 107 metres/year (Delaney and Pollard 1981) and the lateral growth rate of a laccolith is assumed by Corry (1988) to be 200–300 metres/year in maximum. According to other authors (e.g. Scaillet et al. 1995; Hogan and Gilbert 1995; Cruden 1998; Petford and Clemens 2000) the construction of plutons of 102 to 104 km3 takes 102 to 106 years. All these figures for rates of magma ascent through dykes and the formation of lensoidal, tabular plutons are in good agreement with the model results presented here.

In conclusion it can be stated that our model shows that roof uplift and floor depression are viable processes to form tabular plutons, but that floor depression might choke a horizontal and layered magma source, thus narrowing the time span for the build-up of a granitoid lopolith or phacolith.

# ACKNOWLEDGEMENTS

The experiment described here was carried out during an international graduate course about pluton emplacement at Uppsala University in September 2006 with participants from Germany, Greece, Finland and Sweden. Therefore we would like to thank the students of this class – Z. Chemia, M. Dümmler, D. Geis, R. Kraus, T. Kravtsov, T. Laurila, M. Mertineit, R. Nyman, H. Oksanen, E. Piispa, M.D. Tranos, R. Unverricht, J. Woodard, K. Zarins – for building, running and discussing the model with us. Thanks are also due to J.L. Vigneresse and P. Závada for their helpful reviews which greatly improved the manuscript. HAK was funded by the Swedish Research Council. We are grateful to O. Jungmann for improving the figures.

# REFERENCES

1.    Bussels MA (1976) Fracture control of high-level plutonic contacts in the Coastal Batholith of Peru. Proceed Geol Assoc 87: 237–246

2.    Bronstein IN, Semendjajew KA (1987) Taschenbuch der Mathematik. Teubner, Leipzig, pp 1–840

3.    Clemens JD, Petford N (1999) Granitic melt viscosity and silicic magma dynamics in contrasting tectonic settings. J Geol Soc, London 156: 1057–1060

4.    Corry CE (1988) Laccoliths: Mechanics of Emplacement and Growth. GSA Spec Paper 220, Boulder, Colorado, pp 1–110

5.    Cruden AR (1998) On the emplacement of tabular granites. J Geol Soc, London 155: 853–862

6.    Cruden AR, TobischOT , Launeau P (1999) Magnetic fabric evidence for conduit-fed emplacement of a tabular intrusion: Dinkey Creek pluton, central Sierra Nevada batholith, California. J Geophys Res 104: 10 511–10 530

7.    Dehls JF, Cruden AR, Vigneresse JL (1998) Fracture control of late Archean pluton emplacement in the northern Slave Province, Canada. J Struct Geol 20: 1145–1154

8.    Delaney PT, Pollard, DD (1981) Deformation of host rocks and flow of magma during growth of minette dikes and breccia bearing intrusions near Ship Rock, New Mexico. USGS Prof Paper 1202, pp 1–61

9.    Dingwell, DB, Hess, KU, Romano, C (2000) Viscosities of granitic (sensu lato) melts: influence of the anorthite component. Amer Miner 85: 1342–1348

10.   Garde AA, Hamilton MA, Chadwick B, Grocott J, Mc- Caffrey KJW (2002) The Ketilidian orogen of South Greenland: geochronology, tectonics, magmatism and fore-arc accretion during Palaeoproterozoic oblique convergence. Can J Earth Sci 39: 765–793

11.   Grocott J, Garde AA, Chadwick B, Cruden AR, Swager C (1999) Emplacement of rapakivi granite and syenite by floor depression and roof uplift in the Palaeoproterozoic Ketilidian orogen, South Greenland. J Geol Soc, London 156: 15–24

12.   Hack er DB, Petronis MS, Holm DK, Geissma n JW (2007) Shallow level emplacement mechanisms of the Miocene Iron Axis Laccolith Group, Southwest Utah. GSA Rocky Mountain Section Annual Meeting Field Guide, pp 1–49

13.   Hamma rstrom JM, Zen E (1986): Aluminium in hornblende: an empirical igneous geobarometer. Amer Miner 71: 1297–1313

14.   Hecht L, Vigneresse JL (1999) A multidisciplinary approach combining geochemical, gravity and structural data: implications for pluton emplacement and zonation. In: Castro A, Fernandez C, Vigneresse JL (eds) Understanding Granites: Integrating New and Classical Techniques. Geol Soc London Spec Publ 168: pp 95–110.

15.   Hogan JP, Gilbert MC (1995) The A-type Mount Scott Granite sheet: importance of crustal magma traps. J Geophys Res 100: 15 779–15 792

16.   Koyi HA (1991) Mushroom diapirs penetrating into high viscous overburden. Geology 19: 1229–1232

17.   Koyi HA (1997) Analogue modelling; from a qualitative to a

quantitative technique, a historical outline. J Petrol Geol 20: 223–238

18. Koyi HA, Milnes AG, Schmeling H, Talbot CJ, Juhlin C, Zeyen H (1999) Numerical models of ductile rebound of crustal roots beneath mountain belts. Geophys J Int 139: 556–562

19. McCaffrey KJW , Petford N (1997) Are granitic intrusions scale invariant? J Geol Soc, London 154: 1–4

20. O'Brien P (2000) The fundamental Variscan problem: high–temperature metamorphism at different depths and high-pressure metamorphism at different temperatures. In: Franke W, Haak V, Oncken O, Tanner D (eds) Orogenic Processes: Quantification and Modelling in the Variscan Belt. Geol Soc London Spec Publ 179: 369–386

21. Petford N, Clemens JD (2000) Granites are not diapiric! Geol Today 16: 180–184

22. Petford N, Cruden AR, McCaffrey KJW, Vigneresse JL (2000) Granite magma formation, transport and emplacement in the Earth's crust. Nature 408: 669–673

23. Ramberg, H (1981) Gravity, Deformation and the Earth's crust. 2nd edition. Academic Press, London, pp 1–452

24. Rosenberg CL, Berger A, Schmid SM (1995) Observations from the floor of a granitoid pluton: inferences on the driving force of final emplacement. Geology 23: 443–446

25. Scaillet B, Pecher A, Rochette P, Champenois M (1995) The Gangotri granite (Garhwal Himalaya): laccolithic emplacement in an extending collisional belt. J Geophys Res 100: 585–608

26. Weyerma rs, R (1997) Principles of Rock Mechanics. Alboran Science Publishing, Amsterdam, pp 1–360

27. Zimm erma n, NM (2005) Host rock fracture analysis: applying the deformation mechanics associated with shallow igneous intrusion to the fracture bridging theory, McKinney Hills Laccolith, Big Bend National Park. Master's thesis, Texas Tech University, Lubbock. Available electronically from http://hdl.handle.net/2346/1286

# Three Dimensional Modeling of Laterally Loaded Pile Groups Resting in Sand

Amr Farouk Elhakim[a,] , Mohamed Abd Allah El Khouly[a],
and Ramy Awad[b]

[a] Public Works Department, Faculty of Engineering, Cairo University, Egypt
[b] Civil Engineering Department, Queen's University, Canada

## ABSTRACT

Many structures often carry lateral loads due to earth pressure, wind, earthquakes, wave action and ship impact. The accurate predictions of the load–displacement response of the pile group as well as the straining actions are needed for a safe and economic design. Most research focused on the behavior of laterally loaded single piles though piles are most frequently used in groups. Soil is modeled as an elastic-perfectly plastic model using the Mohr–Coulomb constitutive model. The three-dimensional Plaxis model is validated using load–displacement results from centrifuge tests of laterally loaded piles embedded in sand. This study utilizes three dimensional finite element modeling to better understand the main parameters that affect the response of laterally loaded pile groups (2 × 2 and 3 × 3 pile configurations) including sand relative density, pile spacing ($s$ = 2.5 D, 5 D and 8 D) and pile location within the group. The fixity of the pile head affects its load–displacement under lateral loading. Typically, the pile head may be unrestrained (free) head as the pile head is allowed to rotate, or restrained (fixed) head condition where no pile head rotation is permitted. The analyses were performed for both free and fixed head conditions.

# INTRODUCTION

Pile foundations are often necessary to support structures when soil conditions are not favorable, so the pile foundations are required to carry the superimposed load and transfer it to a higher resistant stratum through its bearing and shaft resistance. Piles may also be subjected to lateral loading on abutments and piers which may be caused by earth pressure, ship mooring and berthing forces and wave action. Also lateral loads may occur as a result of wind and unpredicted events such as earthquakes, slope failure and lateral ground spreading induced by liquefaction.

Full scale load tests are the best means for investigating the behavior of laterally loaded piles, they are expensive and time consuming. Other experimental models for determining the movement of the pile under lateral load include 1 g models [1] and [2], and centrifuge tests [3] and [4]. However such methods are only suitable for research purposes because it is impossible to replicate the exact site conditions.

In this study, three dimensional finite element analyses are used to better understand the response of laterally loaded pile groups embedded in sand as outlined in this paper.

# NUMERICAL MODEL

The three-dimensional finite element program Plaxis 3D Foundation was adopted in the numerical analysis of this study. Soil is modeled using quadratic 15-node wedge elements. Piles are modeled using beams which are structural objects used to model slender structures in the ground with a significant flexure rigidity (bending) and a normal stiffness. A three-dimensional mesh is automatically generated taking into account the soil stratigraphy and structure levels as defined by the user. Fig. 1 shows a typical mesh generated by Plaxis 3D Foundation.

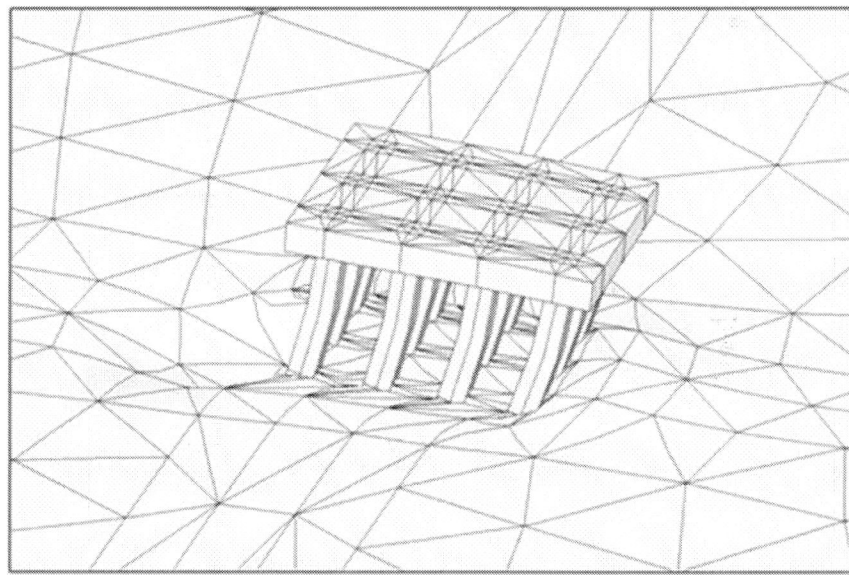

**Figure 1**: Representative mesh for a pile group subjected to lateral loading generated using Plaxis 3D Foundation.

Soil is modeled as an elastic-perfectly plastic model using the Mohr–Coulomb constitutive model built in PLAXIS. In general stress state, the model's stress–strain behaves linearly in the elastic range, with two defining parameters from Hooke's law (Young's modulus, $E$ and Poisson's ratio, ). Failure criteria are defined by the angle of shearing resistance ($\varphi$) and cohesion ($c$).

# MODEL VALIDATION

The three-dimensional Plaxis model is validated using load–displacement results from centrifuge tests of laterally loaded piles embedded in sand [5]. The validation is used to confirm the ability of Plaxis 3D Foundation finite element software to predict the load–displacement relationships for pile groups with various configurations (3 × 3 → 3 × 7).

The centrifuge test is performed on solid square bars having a width of 9.525 mm and an overall length of 304.8 mm (L/B = 32). Each pile is equipped with strain gauges to determine the load distribution within the pile group. The centrifuge test is performed at acceleration equal to 45 g. The equivalent prototype pile width and length are 0.429 and 13.7 m, respectively. The spacing between the piles is three times the pile width. Table 1 summarizes the sand properties used in the centrifuge test and the numerical model [3] and [5]. Fig. 2 and Fig. 3 show the load versus displacement curves of individual piles in laterally loaded pile groups (3 × 3, 3 × 4, 3 × 5, 3 × 6 and 3 × 7) embedded in both sand 1 and sand 2, respectively. The predicted curves (shown as lines) are also found in good agreement with the measured ones (represented by dots). An average error of 10% and a maximum error of 12% are obtained in load component of load–displacement relationship at maximum applied load when the predicted results are compared with the measured ones for all configurations.

**Table 1**: Summary of test sand properties

| Properties | Sand 1 | Sand 2 |
|---|---|---|
| $G_s$ | 2.645 | 2.645 |
| (kN/m³) | 14.5 | 14.05 |
| $D_r\%$ | 55 | 36 |
| ° | 37 | 35 |
| $E_{soil}$ (MPa) | 40–45 | 25–35 |

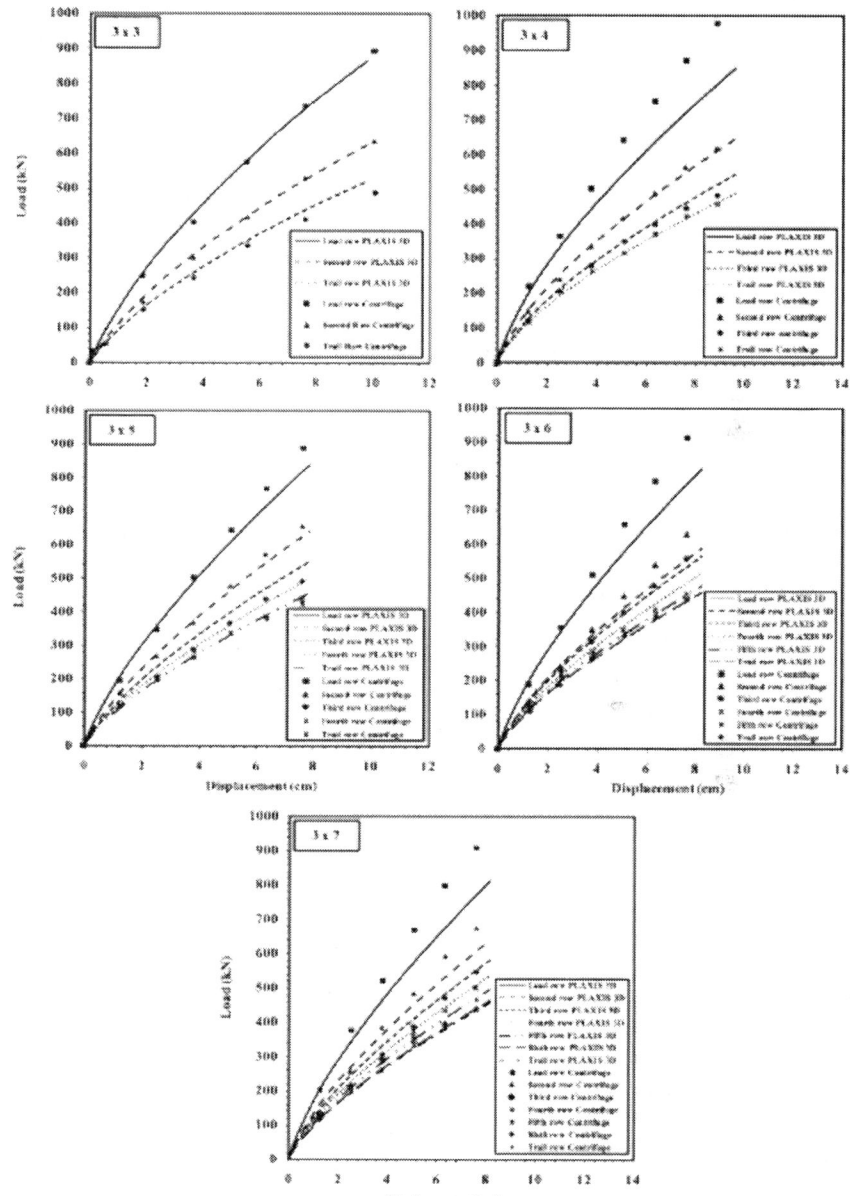

**Figure 2:** Load–displacement curves of laterally loaded pile groups embedded in sand 1.

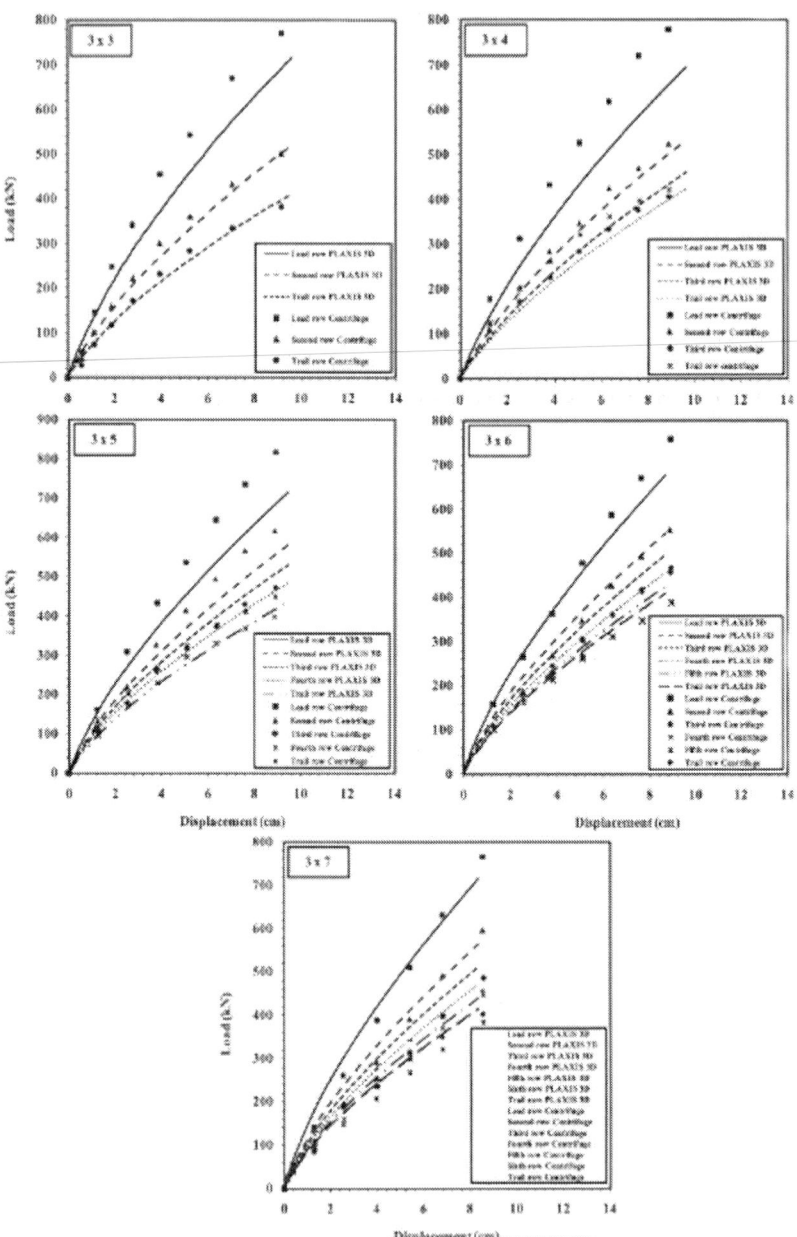

**Figure 3:** Load–displacement curves of laterally loaded pile groups embedded in sand 2.

# PARAMETRIC STUDY

A parametric study is performed to evaluate the effect of sand relative density (angle of shearing resistance and soil modulus), spacing between piles and row location on the behavior of laterally loaded piles in groups embedded in sand. The effect of the investigated parameters on the load–displacement response of the loaded piles and the load distribution within pile groups with fixed head conditions are investigated.

The study investigates the behavior of 15 m long, 0.8 m diameter flexible circular piles. The modulus of elasticity of the pile is taken as 20 GPa which is typical to concrete with $f_{cu}$ = 25 MPa. Pile rigidity can be evaluated following the criteria presented in the Egyptian code of practice [6] where piles are considered rigid when $L/t < 2$ and are considered flexible when $L/t > 4$ where $L$ is the pile length and $t$ is the elastic length calculated according to Eq. (1).

equation(1).

$$t = 5\sqrt{\frac{E_p I_p}{n}} \qquad (1)$$

where:

$E_p$ = Elastic modulus of the pile material (kPa).

$I_p$ = Moment of inertia of the pile (m⁴).

$n$ = Empirical value which depends on the sand relative density.

In the current study, the values of $L/t$ are 6 and 7.78 for piles embedded in loose and dense sands, respectively. Thus piles are considered elastic in different cases. For piles with free head condition, a lateral load of 400 kN is applied at the pile head. For fixed head piles, a total load equal to 400 kN multiplied by the number of piles in the group is applied at the top elevation of the piles. Fig. 4 shows the 2 × 2 and 3 × 3 group configurations used in the analyses.

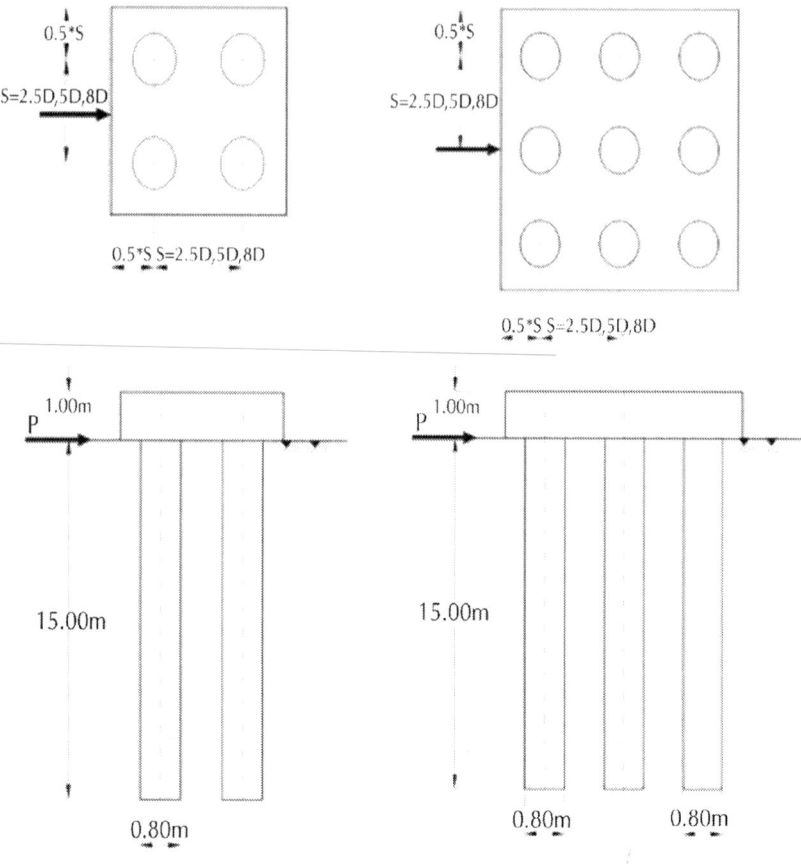

**Figure 4**: Modeled 2 × 2 and 3 × 3 pile group configurations.

Piles are very rarely isolated but are put in groups in order to increase load resistance. Despite the increase in lateral pile group resistance, the individual pile resistance decreases. Each pile pushes the soil behind it creating a shear zone in the soil. These shear zones begin to enlarge and overlap as the lateral load increases especially for the case of closely spaced pile groups. An overlap that occurs between piles in the same row is called "edge effect", while an overlap that occurs between piles in different rows is called "shadowing effect". This overlap in shear zones weakens the soil and results in less lateral resistance per pile. The overlap in shear zones that occurs within a pile group is shown in Fig. 5.

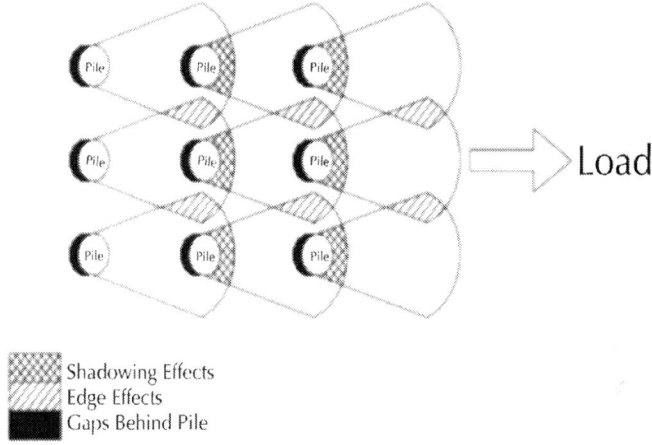

**Figure 5**: Illustration of shadowing and edge effects in a laterally loaded pile group [7].

The effect of varying the relative density, spacing between piles and the effect of row location are investigated. Soil properties used in the analyses are summarized in Table 2. Analyses were performed for both free and fixed pile head conditions for single piles, 2 × 2 and 3 × 3 pile groups. Pile spacings of 2.5, 5 and 8 times the pile diameter were considered. A series of simulations varying the above parameters was performed to investigate the effects of pile to pile spacing and pile head fixation on the load–displacement response of the pile head for piles embedded in loose and dense sands.

**Table 2**: Investigated soil parameters

| Soil formation | Angle of shearing resistance, $\varphi$ (°) | Young's modulus $E$ (MPa) | Poisson's ratio ( ) |
|---|---|---|---|
| Loose | 30 | 20 | 0.3 |
| Dense | 40 | 40 | 0.3 |

# EFFECT OF PILE POSITION AND SPACING BETWEEN PILES

The results of the load–deflection curves of the 2 × 2 free head pile groups embedded in loose and dense sands are shown in Fig. 6. It is observed that a rear row pile deflects more than a front row pile for any given spacing. Also for a given row, the lateral deflection decreases with the increase of pile spacing. The load–deflection response for a pile group is softer than that of a single pile. The difference becomes less noticeable as the pile spacing increases. For $S$ = 8 D, the simulated load–deflection curves approach the load–deflection response of a single pile for both loose and dense sands with a difference of 13% and 3% for loose and dense sands, respectively. This agrees with the findings of Rollins [8] and Krishnamoorthy and Sharma [9]. The results also indicate that the shadowing effect diminishes at a faster rate in dense sand compared to loose sand. A similar load–deflection response is noticed for 3 × 3 pile groups as shown in Fig. 7.

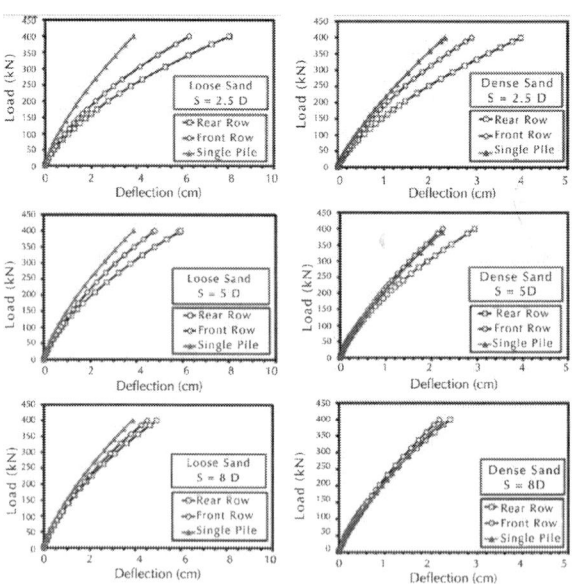

**Figure 6:** Load versus deflection of 2 × 2 free headed pile groups in loose and dense sand.

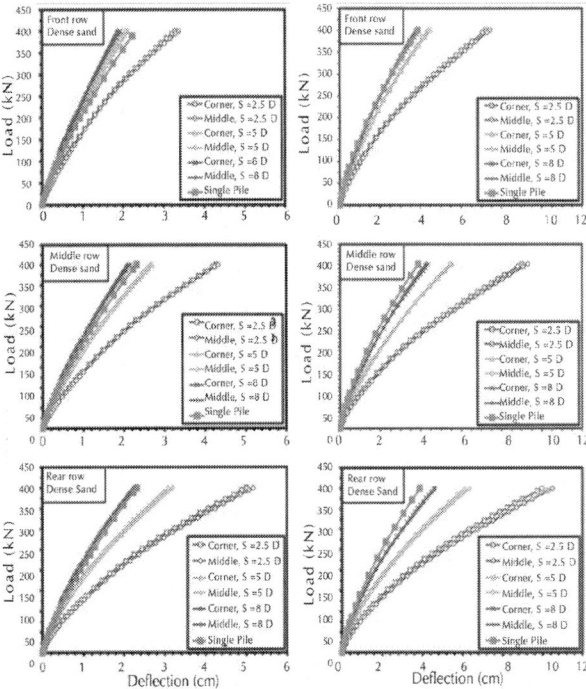

**Figure 7**: Load versus deflection of 3 × 3 free headed pile groups in loose and dense sands.

To allow a better perspective, a normalization procedure [10] is adopted to represent the results in a non-dimensional format for application to piles of varying diameters under different loads. The pile spacing is normalized by pile diameter (*S/D*) and the deflection is normalized through the following procedure:

equation(2)

$$I_{xx} = \left(U_{xx} * E_{soil} * D\right) / Load\, Per\, Pile \qquad (2)$$

where:

$I_{xx}$ = Normalized deflection at pile top.

$U_{xx}$ = Lateral deflection.

$E_{soil}$ = Soil modulus.

$D$ = Pile diameter.

Fig. 8a and b show the normalized deflection at pile top versus normalized spacing for 2 × 2 and 3 × 3 free headed pile groups, respectively. It is noted that as the pile spacing increases, deflections approach a constant value in dense sands which indicates that the shadowing effect is minimized. For piles embedded in loose sand, lateral deflections do not reach a constant value with the increase in pile spacing for the 2 × 2 pile group. For the 3 × 3 pile group in dense sand, lateral deflections decrease with the increase in pile spacing up to a normalized spacing of approximately 5 at which the movement becomes constant. For 3 × 3 pile groups embedded in loose sand, the rate of lateral movement of the 3 × 3 pile group decreases for normalized pile spacing greater than 5 compared to smaller spacing but does not reach a constant value. Thus, it may be concluded that the shadowing effect diminishes at a greater rate in dense sand compared to loose sand.

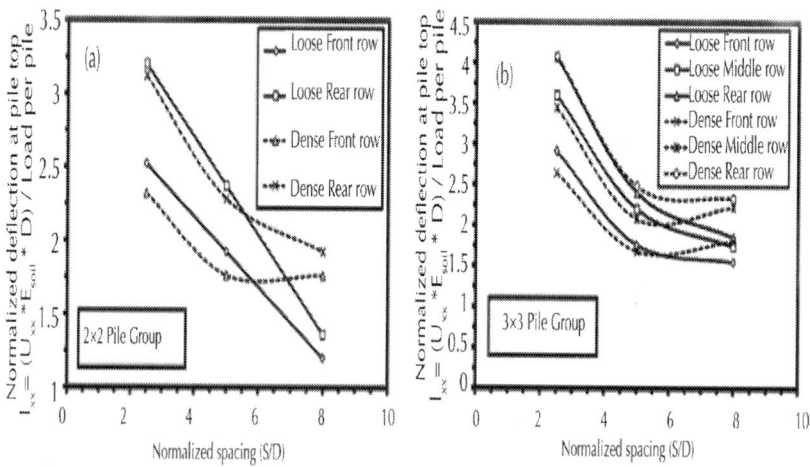

**Figure 8**: Normalized deflection at pile top versus normalized spacing for (a) 2 × 2 (b) 3 × 3 free headed pile groups.

# EFFECT OF PILE HEAD FIXATION

The fixity of the pile head affects its load–displacement under lateral loading. Typically, the pile head may be unrestrained (free) head as the pile head is allowed to rotate or restrained (fixed) head condition where no pile head rotation is permitted. This section discusses the effect of restricting the rotation of the pile head on the load–displacement response of laterally loaded pile groups embedded in loose and dense sands. For piles with free head condition, a lateral load of 400 kN is applied at the pile head. For fixed head piles, a total load equal to 400 kN multiplied by the number of piles in the group is applied at the top elevation of the piles.

Fig. 9 shows the applied load per pile normalized to the average load per pile versus the center to center pile spacing normalized to the pile diameter for 3 × 3 capped piles embedded in loose sand. The front row in the group carries the highest load, middle and back row piles carry smaller loads for a given displacement. These results are in agreement with the findings made by Ruesta and Townsend [11], Rollins et al. [12] and Rollins[13]. Furthermore, it is observed that as the pile spacing increases, the normalized lateral load approaches unity since the shadowing effect decreases with the increase in pile to pile spacing in agreement with Zhang and Small [10].

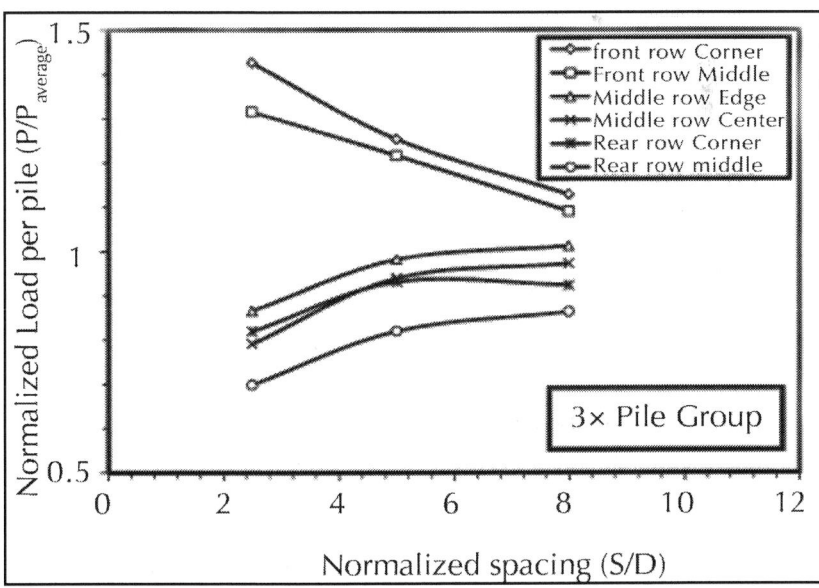

**Figure 9**: Normalized load per pile versus the normalized pile center to center spacing for loose sand.

Simulations of piles embedded in loose and dense sands were performed. Three center-to-center pile spacings were considered in the analyses (2.5 D, 5 D and 8 D). The results of these simulations for piles embedded in loose sand are shown in Fig. 10. As expected, the pile lateral head movement decreases for fixed head condition compared to free head condition in agreement with the findings of Duncan et al. [14]. The ratio of the maximum pile head movements under free and fixed conditions varied between approximately 1.3 and 2.4 depending on the pile position and spacing. Similar findings were found for piles embedded in loose sands placed in 2 × 2 groups as well as piles placed in dense sands with results presented by Awad [15].

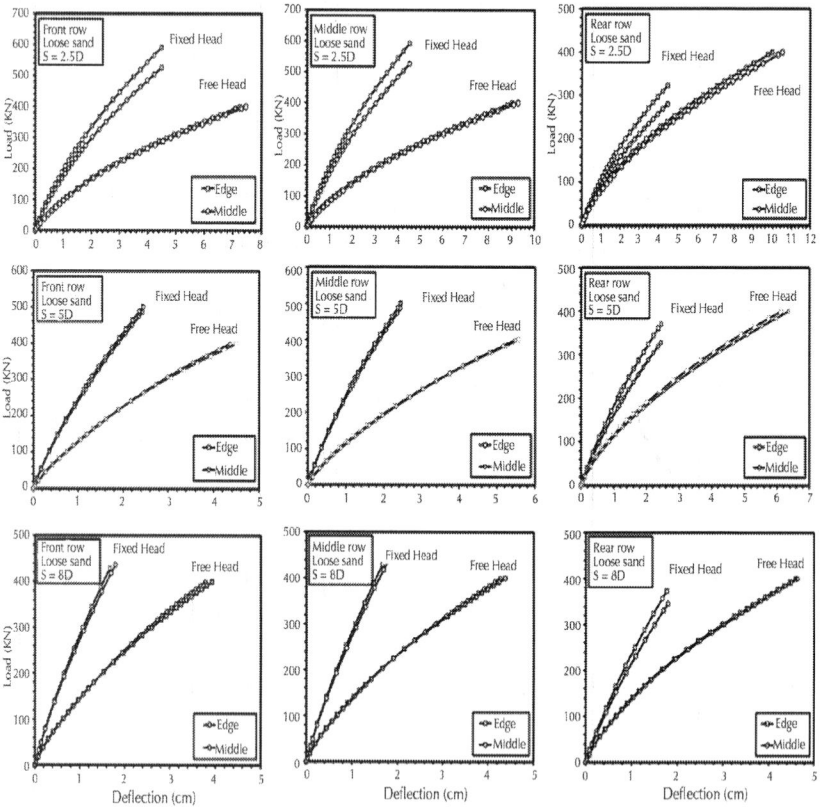

**Figure 10**: Effect of head fixation on the load–deflection response of piles embedded in loose sand at spacings of 2.5 D, 5 D and 8 D.

# SUMMARY AND CONCLUSIONS

Three dimensional finite element modeling is used to assess the load-movement response of laterally loaded piles embedded in sand. The three-dimensional Plaxis model is validated using results from centrifuge tests of laterally loaded piles embedded in sand. The validation is used to confirm the ability of Plaxis 3D Foundation finite element software to predict the load–displacement relationships for pile groups with various configurations.

A parametric study is performed to evaluate the effect of sand relative density (angle of shearing resistance and soil modulus), spacing between piles and row location on the behavior of laterally loaded piles in groups embedded in loose and dense sand formations. Based on the results of the parametric study, the following conclusions are made.

1.  It is concluded that piles installed in groups at close spacings deflect more than a single pile subjected to the same lateral load per pile because of group effect.

2.  The load deflection response of piles in groups is softer compared to single pile load deflection curve. The difference becomes less significant with the increase in pile spacing due to the decrease of the shadowing effect.

3.  For pile groups with fixed head, the front row in the group carries the highest load, middle and back row piles carry smaller loads for a given displacement. The ratio of the maximum pile head movements under free and fixed conditions varied between approximately 1.3 and 2.4 depending on the pile position and spacing.

4.  The shadowing effect diminishes at a greater rate in dense sand compared to loose sand.

# REFERENCES

1.  S. Prakash, Behavior of pile groups subjected to lateral loads (Ph.D. dissertation), University of Illinois at Urbana, IL, USA, 1962.

2.  N.R. Patra, P.J. Pise Ultimate lateral resistance of pile groups in sand ASCE J. Geotech. Geoenviron. Eng., 127 (6) (2001), pp. 481–487

3.  M. McVay, D. Bloomquist, D. Vanderlinde, J. Clausen Centrifuge modeling of laterally loaded pile groups in sands ASTM Geotech. Test. J., 17 (1994), pp. 129–137

4.  L.G. Kong, L.M. Zhang Rate-controlled lateral-load pile tests using a robotic manipulator in centrifuge ASTM Geotech. Test. J., 30 (3) (2007), pp. 192–201

5.  M. McVay, R. Casper, T. Shang Lateral response of three-row groups in loose to dense sands at 3D and 5D pile spacing ASCE J. Geotech. Eng., 121 (5) (1995), pp. 436–441

6.  Egyptian Code of Practice for Soil Mechanics, Design and Construction of Foundations, Housing and Building Research Center, Giza, Egypt, 2001.

7.  R.L. Mokwa, Investigation of the resistance of pile caps to lateral loading (Doctoral dissertation), Virginia Polytechnic Institute and State University, Virginia, 1999.

8.  K.M. Rollins, Static and dynamic lateral load behavior of pile groups based on full-scale testing, in: Proceedings of the Thirteenth International Offshore and Polar Engineering Conference Honolulu, Hawaii, USA, 2003, 2, pp. 506–513.

9.  Krishnamoorthy, K.J. Sharma, Analysis of single and group of piles subjected to lateral load using finite element method, in: Proceedings of the 12th International Conference of International Association for Computer Methods and Advances in Geomechanics (IACMAG) Goa, India, 2008, pp. 3111–3116.

10. H. Zhang, J. Small Analysis of capped pile groups subjected to horizontal and vertical loads Comput. Geotech., 26 (1999), pp. 1–21

11. P.F. Ruesta, F.C. Townsend Evaluation of laterally loaded pile group at roosevelt bridge ASCE J. Geotech. Geoenviron. Eng., 123 (12) (1997), pp. 1153–1162

12. K.P. Rollins, K.T. Peterson, T.J. Weaver Lateral load behavior of full-scale pile group in clay ASCE J. Geotech. Geoenviron. Eng., 124 (6) (1998), pp. 468–478

13. [13] K.M. Rollins Static and Dynamic Lateral Load Behavior of Pile Groups Based on Full-Scale Testing Master of Science, Brigham Young University, Utah, USA (2003)

14. J.M. Duncan, L.T. Evans, P.S. Ooi Lateral load analysis of single piles and drilled shafts ASCE J. Geotech. Eng., 120 (6) (1994), pp. 1018–1033

15. R. Awad, Three dimensional finite element modeling of laterally loaded pile groups in sand (MSc thesis), Department of Public Works, Cairo University, 2011

# Assessing the Impact of Ground-Motion Variability and Uncertainty on Empirical Fragility Curves

Ioanna Ioannou[a], John Douglas[b], and Tiziana Rossetto[a]

[a]EPICentre, Civil, Environmental and Geomatic Engineering, University College London, Gower Street, London WC1E 6BT, United Kingdom
[b]BRGM – DRP/RSV, 3 avenue C. Guillemin, BP 36009, 45060 Orleans Cedex 2, France

## ABSTRACT

Empirical fragility curves, constructed from databases of thousands of building-damage observations, are commonly used for earthquake risk assessments, particularly in Europe and Japan, where building stocks are often difficult to model analytically (e.g. old masonry structures or timber dwellings). Curves from different studies, however, display considerable differences, which lead to high uncertainty in the assessed seismic risk. One potential reason for this dispersion is the almost universal neglect of the spatial variability in ground motions and the epistemic uncertainty in ground-motion prediction. In this paper, databases of building damage are simulated using ground-motion fields that take account of spatial variability and a known fragility curve. These databases are then inverted, applying a standard approach for the derivation of empirical fragility curves, and the difference with the known curve is studied. A parametric analysis is conducted to investigate the impact of various assumptions on the results. By this approach, it is concluded that ground-motion variability leads to flatter fragility curves and that the epistemic uncertainty in the ground-motion prediction equation used can have a dramatic impact on the derived curves. Without dense ground-motion recording networks in the epicentral area empirical curves will remain highly uncertain. Moreover, the use of aggregated damage observations appears to substantially increase uncertainty in the empirical fragility assessment. In contrast, the use of limited randomly-chosen un-aggregated samples in the affected area can result in good predictions of fragility.

# INTRODUCTION

Fragility curves of buildings exposed to earthquakes express the likelihood of damage to these assets from future seismic events. Empirical fragility curves are based on the statistical analysis of post-earthquake observations of the damage sustained by the exposed buildings and the corresponding ground-motion intensity level at the building locations. Currently at least 119 empirical fragility curves have been published [1]. These curves have generally been constructed assuming that the measurement error in the intensity-measure levels (IMLs) is negligible. However, given the general lack of a dense strong-motion network in the areas of damaging earthquakes, the intensity levels are typically estimated though ground motion prediction equations (GMPEs) or, more recently, ShakeMaps. Hence, the IMLs are associated with high measurement error. In recent years, a handful of studies have proposed undertaking a Bayesian regression analysis to explicitly model this error [2], [3] and [4]. Nonetheless, the impact of this measurement error on empirical fragility curves is not well understood.

This study aims to examine the impact of the measurement error in the IMLs on empirical fragility curves. A simulation study is undertaken to investigate this issue, following a similar philosophy to Gehl et al. [5], who studied the influence of the number of dynamic runs on the accuracy of fragility curves. In the next section the method of simulation is introduced. This approach is applied in the subsequent section to undertake a parametric analysis to study the influence of different assumptions on the empirical fragility curves. The paper finishes with some discussion of the results, the limitations of existing empirical fragility curves, implications for the development of future empirical fragility functions as well as possible ways forward.

# METHOD

The impact of ground-motion variability and uncertainty on empirical fragility curves is studied here by undertaking a series of experiments. In these, an earthquake with specified characteristics (i.e. magnitude, location and faulting mechanism) affects a number of buildings ($N_{Buildings}$) located in a number of towns ($N_{Towns}$).

The construction of empirical fragility curves requires observations of two variables, namely: the damage sustained by the considered buildings and their corresponding IMLs. IMLs are generated assuming the absence or the presence of ground-motion observations.

## Seismic Damage

In this study, the damage experienced by each building in the affected area is considered random due to the uncertainty in its IML as well as the uncertainty in its structural performance given this IML. Therefore, seismic damage for each building is determined here by modelling these two uncertainties through a Monte Carlo analysis. The procedure adopted is an extension of the procedure used by Douglas [6] in order to study the density of seismic networks required to monitor ground motions from induced seismicity. According to this analysis, a large number, $N_{Realisations}$, of IMLs and subsequent damage states are generated.

According to the procedure proposed by Douglas [6], each realisation of IMLs for the considered buildings occurs from the generation of a ground-motion field using a given GMPE coupled with models of spatial variability. To simulate the spatially-correlated ground-motion fields the procedure of Strasser and Bommer ([7], pp. 2625–2626) is used. The package geoR [8] of the statistical software R allows such fields to be generated quickly and then manipulated. The between-event and within-event ground-motion variabilities are included within the fields. The deterministic ground-motion field produced by evaluating the considered IMLs for all building locations in the region is perturbed by the addition of a random field derived from a multivariate normal distribution based on a standard deviation equal to the within-event variability of the selected GMPE and an exponential correlation function, G (h), which is found to fit the observed spatial correlation of earthquake ground motions [9] and [10]:

$$G(h) = \exp\left(-\frac{h}{h_0}\right)$$

(1)

where h is the separation distance between locations of interest and $h_0$ is the correlation range. Because one ground-motion field differs greatly from another, this procedure is repeated many times so that robust conclusions can be drawn from the combined results. The sensitivity of the results on the chosen GMPE, the value of $h_0$ and other input parameters (e.g. size of the region, density of ground-motion measurements and aggregation level) are investigated in this paper.

In order to simulate earthquake-damage fields, a known fragility curve expressing the fragility of hypothetical buildings in a region is applied. This curve takes as input the simulated ground-motion fields and yields the building damage observations used as the empirical dataset for the study. Consequently, the impact of sparse or uncertain observations on fragility curves can be evaluated by comparing the resulting empirical fragility curves derived from different sampling and assumptions, with the curve used as input in the simulations. The advantage of this approach is that the 'true' fragility of the structures is known and can be compared with the empirical fragility curves resulting from the experiments.

In particular, for a realisation k, resulting in $iml_{realisationk}$, the damage sustained by each building is randomly generated as follows. In order to simplify the analysis, the determination of the exact damage state of each building is not required. Instead, we concentrate on whether the building has reached or not a given damage state, $ds_i$, assuming an appropriate fragility curve from the literature. In particular, for a realisation k, the building, j, is assigned an indicator, $Y_{jk}$, where:

$$Y_{jk} = \begin{cases} 1 & DS \geq ds_i \\ 0 & DS < ds_i \end{cases} \qquad (2)$$

The indicator is randomly assigned to the building j, by assuming that it follows a special case of the binomial distribution, termed the Bernoulli distribution:

$$Y_{jk}\left|IM_{true} = iml_{true',k} \sim \binom{n}{y_{jk}} \mu_{jk}{}^{y_{jk}} \left[1 - \mu_{jk}\right]^{n-y_{jk}}\right.$$

$$\text{where } \mu_{jk} = P(DS \geq ds_i \left|iml_{true',k}\right.) = \Phi\left(\frac{\ln(iml_{true',k}) - \lambda_k}{\zeta_k}\right) \qquad (3)$$

where n is the number of buildings for a given intensity measure level, $iml_{realisationk'}$, and in this case, n=1; $\mu_j$ equals the probability that the building is in damage state $ds_i$ or above given $iml_{realisationk'}$; $\mu_{jk}$ is the mean of the Bernoulli distribution, which is typically expressed in the literature in terms of a cumulative lognormal distribution;   is the cumulative standard normal distribution;   is the lognormal mean; and   is the lognormal standard deviation.

# Ground-motion Intensity

The determination of the IML at the location of each building is necessary for the construction of empirical fragility curves. These levels are considered known and measured without uncertainty, an assumption commonly made when deriving such curves. The determination of these 'true' IMLs depends on the absence or presence of ground-motion observations.

## The Absence of Ground Motion Recording Stations

In the absence of ground-motion records, empirical fragility curves are derived here by following the common assumption that the IML at the location of each building is equal to the median values obtained from a pre-selected GMPE. It should be noted that the selected equation is not necessarily the same as the one used to generate the damage levels because in practice the appropriate GMPE for an earthquake is not known.

## *The Presence of Ground Motion Recording Stations*

The random fields of peak ground accelerations (PGAs) are recovered assuming the presence of ground motion recording stations located at a varying number of buildings. This consideration suggests that the IMLs for the buildings at which records are available are known and equal to the corresponding values provided by the random field. The IMLs for the remaining buildings are estimated from these records using a procedure known as kriging. In this study, kriging uses the same correlation model as the one used for the generation of the random fields.

## Empirical Fragility Curves

The empirical fragility curve is then constructed for the k realisations of IMLs by considering the Yjkindicators generated for all considered buildings, according to the procedure described in Section 2.1, and the corresponding 'true' IMLs as determined in Section 2.2. Their construction follows the procedure proposed in the Global Earthquake Model empirical fragility assessment guidelines [3]:

$$Y_{jk}\left|IM_{true} = iml_{'true',k} \sim \binom{n}{y_{jk}}\mu_{jk}^{y_{jk}}\left[1-\mu_{jk}\right]^{n-y_{jk}}\right.$$

$$\text{where } \mu_{jk} = P(DS \geq ds_i \left| iml_{'true',k}) = \Phi\left(\frac{\ln(iml_{'true',k}) - \lambda_k}{\zeta_k}\right)\right. \tag{4}$$

where $\lambda_k$ is the lognormal mean and $\zeta_k$ is the lognormal standard deviation for realisation k estimated by Eq. (4). Both of these parameters fully describe the empirical fragility curve for realisation k.

# RESULTS

A simulation study is undertaken to gain insight into the influence on empirical fragility curves of the measurement error in IMLs. Within this study, five hypothetical Turkish towns (termed A to E), distributed at equal intervals of 22.5 km as presented in Fig. 1, are considered. For simplicity, all ground motions are predicted assuming uniform stiff soil (e.g., $Vs_{30}$=400 m/s) conditions throughout the region. Each town includes N=10,000 buildings, spatially distributed on the nodes of a grid with spacing of 20 m. To reduce the complexity of this study, masonry buildings are assumed with fragility expressed by the analytical fragility curves of Erberik [11], which adopt PGA as the intensity measure and use three damage states. Here, only the curve corresponding to moderate damage state is used for simplicity. The buildings in the five towns (i.e. 5 towns×10,000 buildings=50,000 buildings in total) are assumed to be affected by a normal-faulting earthquake with known location and magnitude. The effect on the empirical fragility curves of the uncertainty in PGA estimates is examined by assuming the absence as well as the presence of strong-motion stations in the epicentral region. Table 1 lists the characteristics of the different parametric analyses conducted.

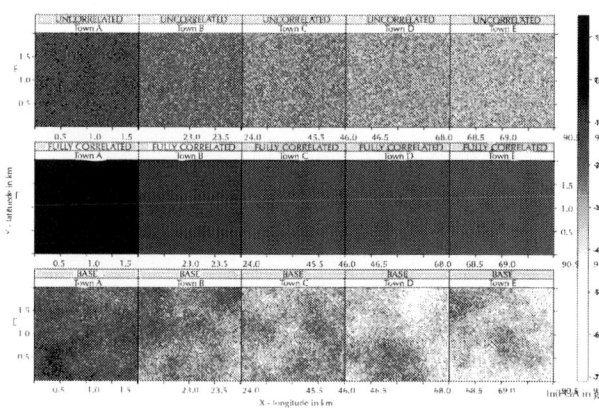

**Figure 1**: Location of the epicentre (star), the five towns and realisations of PGA fields for 'UNCORRELATED' (top row), 'FULLY CORRELATED' (middle row), and 'BASE'(bottom row) scenarios. How the PGA (in g) fields shown here were generated is discussed later on in the paper.

**Table 1:** Scenario simulations and characteristics of the inverted fragility curves

| Name | $M_w$ | $h_0$(km) | Grid (m²) | Aggregation of buildings | N | PCA$_{median}$ | PCA$_{realisation}$ | PGA=0.07 g | | | PGA=0.20 g | | |
|---|---|---|---|---|---|---|---|---|---|---|---|---|---|
| | | | | | | | | 95% | Mean | 5% | 95% | Mean | 5% |
| Sensitivity to uncertainty in GMPE | | | | | | | | | | | | | |
| CHECK | 7.2 | 10 | 20×20 | No | 5×10,000 | GMPE1 | GMPE$_1$ | 9.0 | 9.2 | 9.5 | 67.1 | 67.7 | 68.3 |
| BASE | 7.2 | 10 | 20×20 | No | 5×10,000 | GMPE1 | GMPE$_1$ | 11.0 | 23.0 | 38.0 | 37.6 | 60.7 | 80.6 |
| Sensitivity to GMPE | | | | | | | | | | | | | |
| GMPE2 | 7.2 | 10 | 20×20 | No | 5×10,000 | GMPE2 | GMPE$_1$ | 4.3 | 14.3 | 27.9 | 24.7 | 42.5 | 60.7 |
| GMPE3 | 7.2 | 10 | 20×20 | No | 5×10,000 | GMPE3 | GMPE$_1$ | 3.7 | 13.1 | 26.1 | 23.8 | 41.5 | 59.4 |
| Sensitivity in **h0** | | | | | | | | | | | | | |
| Uncorrelated | 7.2 | Uncorrelated | 20×20 | No | 5×10,000 | GMPE1 | GMPE$_1$ | 8.4 | 18.7 | 32.2 | 44.9 | 62.4 | 78.4 |
| Fully correlated | 7.2 | Fully correlated | 20×20 | No | 5×10,000 | GMPE1 | GMPE$_1$ | 0.07 | 19.8 | 71.0 | 7.6 | 63.2 | 99.0 |
| Sensitivity to buildings' density | | | | | | | | | | | | | |
| Density=50 m×50 m | 7.2 | 10 | 50×50 | No | 5×10,000 | GMPE1 | GMPE$_1$ | 12.0 | 23.2 | 36.5 | 42.7 | 60.2 | 76.5 |
| Sensitivity to the number of buildings | | | | | | | | | | | | | |
| N=5×100 (Ordered 1) | 7.2 | 10 | 20×20 | No (Ordered 1) | 5×100 | GMPE1 | GMPE$_1$ | 7.4 | 22.9 | 43.8 | 29.9 | 60.4 | 87.9 |
| N=5×100 (Ordered 2) | 7.2 | 10 | 20×20 | No (Ordered 2) | 5×100 | GMPE1 | GMPE$_1$ | 8.9 | 22.6 | 39.4 | 35.6 | 61.0 | 83.1 |
| N=5×100 (Random) | 7.2 | 10 | 20×20 | No (Random) | 5×100 | GMPE1 | GMP$_{cl}$ | 9.7 | 22.8 | 38.5 | 36.9 | 60.6 | 81.5 |
| N=5×1000 (Ordered 1) | 7.2 | 10 | 20×20 | No (Ordered 1) | 5×1000 | GMPE1 | GMPE$_1$ | 9.1 | 22.8 | 40.1 | 34.5 | 60.5 | 84.1 |
| Sensitivity to buildings aggregation | | | | | | | | | | | | | |

| Name | $M_w$ | $h_0$(km) | Grid (m²) | Aggregation of buildings | N | $PGA_{median}$ | $PGA_{realisation}$ | PGA=0.07 g 95% | PGA=0.07 g Mean | PGA=0.07 g 5% | PGA=0.20 g 95% | PGA=0.20 g Mean | PGA=0.20 g 5% |
|---|---|---|---|---|---|---|---|---|---|---|---|---|---|
| Aggregated=1 point per town | 7.2 | 10 | 20×20 | 1×1 per town | 5×10,000 | $GMPE_1$ | $GMPE_1$ | 1.7 | 17.7 | 44.1 | 18.8 | 63.4 | 97.7 |
| Aggregated=100 points per town | 7.2 | 10 | 20×20 | 10×10 per town | 5×10,000 | $GMPE_1$ | $GMPE_1$ | 2.1 | 18.0 | 40.8 | 22.5 | 63.6 | 96.5 |
| Kriging | | | | | | | | | | | | | |
| 5×1 st, COR. | 7.2 | 10 | 20×20 | 10×10 p town | 5×100 | GMPE1 | 1 station p town | 12.1 | 25.2 | 39.5 | 28.6 | 48.7 | 73.9 |
| 5×10 st, COR. | 7.2 | 10 | 20×20 | 10×10 p town | 5×100 | GMPE1 | 10 stations p town | 15.5 | 21.5 | 28.4 | 42.1 | 53.9 | 65.2 |
| 5×50 st, COR. | 7.2 | 10 | 20×20 | 10×10 p town | 5×100 | GMPE1 | 50 stations p town | 12.6 | 16.8 | 21.0 | 53.0 | 59.8 | 66.2 |
| 5×100 st, COR. | 7.2 | 10 | 20×20 | 10×10 p town | 5×100 | GMPE1 | 100 stations p town | 5.8 | 9.1 | 12.6 | 61.6 | 68.1 | 75.0 |
| 5×1 st, UNCOR. | 7.2 | Uncorrelated | 20×20 | 10×10 p town | 5×100 | GMPE1 | 1 station p town | 0.0 | 21.1 | 59.8 | 20.8 | 73.6 | 100.0 |
| 5×10 st, UNCOR. | 7.2 | Uncorrelated | 20×20 | 10×10 p town | 5×100 | GMPE1 | 10 stations p town | 7.1 | 20.5 | 31.3 | 57.5 | 73.3 | 89.0 |
| 5×50 st, UNCOR. | 7.2 | Uncorrelated | 20×20 | 10×10 p town | 5×100 | GMPE1 | 50 stations p town | 9.6 | 17.5 | 25.0 | 62.6 | 71.3 | 79.4 |
| 5×100 st, UNCOR. | 7.2 | Uncorrelated | 20×20 | 10×10 p town | 5×100 | GMPE1 | 100 stations p town | 5.8 | 9.1 | 12.6 | 61.7 | 67.9 | 74.6 |
| 5×1 st, FCOR. | 7.2 | Fully correlated | 20×20 | 10×10 p town | 5×100 | GMPE1 | 1 station p town | 11.0 | 25.0 | 40.6 | 27.3 | 48.9 | 73.7 |
| 5×10 st, FCOR. | 7.2 | Fully correlated | 20×20 | 10×10 p town | 5×100 | GMPE1 | 10 stations p town | 15.3 | 21.5 | 27.9 | 44.5 | 54.1 | 63.5 |
| 5×50 st, FCOR. | 7.2 | Fully correlated | 20×20 | 10×10 p town | 5×100 | GMPE1 | 50 stations p town | 12.6 | 16.7 | 21.2 | 52.7 | 59.8 | 66.8 |
| 5×100 st, FCOR. | 7.2 | Fully correlated | 20×20 | 10×10 p town | 5×100 | GMPE1 | 100 stations p town | 5.8 | 9.0 | 12.6 | 61.7 | 67.9 | 74.6 |

Common to all sensitivity analyses is the reference scenario, termed 'BASE', which assumes that the aforementioned 50,000 buildings are affected by an earthquake, with moment magnitude $M_w$=7.2, whose epicentre is located 10 km west of the centre of town A, as depicted in Fig. 1. In addition, the PGAs experienced by each building are estimated by the GMPE (termed GMPE1) derived by Akkar and Bommer [12], which models a large intra-event as well as inter-event variability. Following the procedure outlined in Section 2.1, this variability as well as the spatial correlation is taken into account by generating 1000 realisations of the PGA levels experienced by the 50,000 buildings. It should be noted that we assume $h_0$=10 km, which is a typical estimate of this parameter in recent studies [9] and [10], although it should be noted that this parameter appears to vary with the structural period, geographical location and earthquake and consequently there is much uncertainty over the appropriate value. For each realisation of the PGAs, the damage experienced by each building is generated according to the procedure outlined in Section 2.1. The construction of empirical curves requires the determination of the 'true' PGAs. For this scenario, the absence of strong-motion stations in the vicinity is considered. Therefore, the 'true' values are considered equal to the median GMPE1 values. The empirical fragility curves for moderate damage for each realisation, depicted in Fig. 2a, are substantially flatter than the fragility curves of Erberik [11], which express the actual fragility of the masonry buildings. This suggests that the ground-motion variability leads to a considerable increase in the uncertainty of empirical fragility curves.

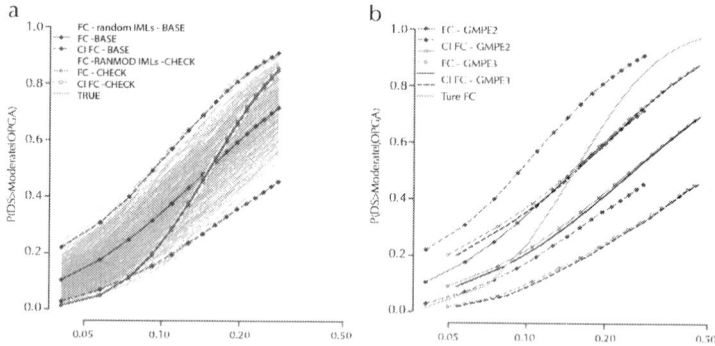

**Figure 2**: Sensitivity of fragility curves (FC) and their 90% confidence intervals (CI FC) to (a) uncertainty in GMPE and (b) selection of GMPE..

The impact of the variability in PGA on the mean empirical fragility curve as well as its 90% confidence intervals is examined here by considering an alternative scenario, termed 'CHECK', which sets the variability in GMPE1 to zero. According to this scenario, the 'true' PGAs are considered known and equal to their corresponding values obtained from the 1000 random fields. The procedure described above to construct empirical fragility curves is again used and the results obtained are presented in Fig. 2a. The mean fragility curve for this scenario is almost identical to the 'true' fragility curve, as expected. The very narrow width of the 90% confidence intervals can be attributed to the large sample size, i.e. 50,000 buildings, used to produce each curve.

The differences in the empirical fragility curves constructed by the two scenarios, i.e. 'BASE' and 'CHECK', are quantified in Table 1, which lists the mean and 90% confidence intervals of these curves for two example PGAs (0.07 g and 0.20 g). The flatter mean curve for 'BASE' leads to a P (DS≥ moderate damage/PGA=0.07 g) that is 150% higher than its 'CHECK' counterpart. The difference is reduced to 10% for 0.20 g. With regard to the width of the 90% confidence intervals, this appears to be 54 and 36 times wider for 'BASE' than 'CHECK' given PGA=0.07 g and 0.20 g, respectively.

Overall, the variability in the PGAs leads to a substantial increase in uncertainty in the empirical fragility curves, which is manifested both in terms of a flatter curve as well as wider confidence intervals.

In the construction of empirical fragility curves for 'BASE', the 'true' IMLs have been considered equal to the median values of GMPE1. Nonetheless, the selection of a GMPE to express the 'true' PGA levels depends on the analyst. What is the impact of using median values of an alternative GMPE? To address this question, the GMPEs proposed by Cauzzi and Faccioli [13] and Zhao et al. [14], identified by Delavaud et al. [15] as suitable for Europe and the Middle East, are used to estimate the 'true' PGAs. The modified 'BASE' scenario is re-run and the estimated mean and 90% confidence intervals for the two scenarios, termed 'GMPE2' and 'GMPE3', respectively, are depicted in Fig. 2b. The 'true' intensity measure levels for 'GMPE2' and 'GMPE3' are higher than their 'BASE' counterparts as presented in Fig. 3. This leads to the mean fragility curves for the two scenarios to be shifted to the right indicating an improved seismic performance than their 'BASE' counterparts (see Fig. 2b). It should be noted, however, that improved performance does not make the curves closer to the 'true' fragility curve [11]. In particular, for PGA=0.07 g, the difference between the mean fragility curve for 'GMPE2' and the 'true' fragility curve is reduced to 55% and 37% for 'GMPE2' and 'GMPE3', respectively. By contrast, the width of the 90% confidence intervals is the same as for 'BASE'. This is expected given that the width is related to the ground-motion variability, which remains the same for all three scenarios (the standard deviations associated to GMPE2 and GMPE3 were assumed equal to that for GMPE1). The true total standard deviations of GMPE2 (0.344 in terms of $\log_{10}$ PGA) and GMPE3 (0.314) are similar to that of GMPE1 (0.279) and hence using these values instead would not have a large impact on the results.

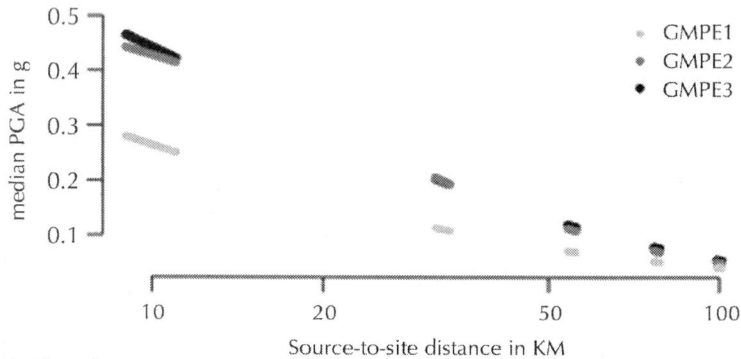

**Figure 3**: Median PGAs (in g) against the source-to-site distance for GMPE1, GMPE2 and GMPE3.

Having established that the event's characteristics mainly affect the range of IMLs, the influence on the empirical fragility curves of the spatial correlation parameters, namely $h_0$ and the building density, are examined next.

The impact on the empirical fragility curves of the correlation introduced by the exponential model, used to model the Gaussian random fields in Section 2.1, is examined here by re-running the 'BASE' scenario assuming that the intra-event residuals in GMPE1 are uncorrelated or fully correlated. According to the former scenario, the intensity measure level experienced by each building does not depend on the intensity at adjacent buildings. Fig. 4a shows that this assumption leads to a steeper mean fragility curve, which appears to be closer to its 'true' counterpart, and notably smaller confidence intervals than 'BASE'. This agrees with the findings of Crowley et al. [9]. In particular for PGA=0.20 g, the confidence intervals around the empirical fragility curves are 22% narrower than for 'BASE'. By contrast, if the intra-event residuals are considered fully correlated, the confidence intervals round the fragility curve are significantly wider. In particular, they are 2.6 times wider than their 'BASE' counterparts (see Table 1).

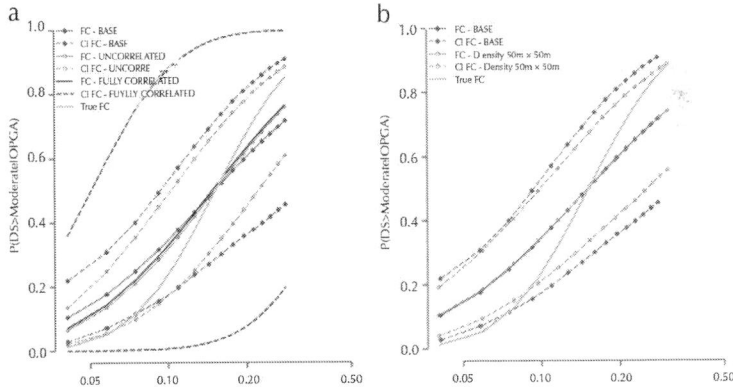

**Figure 4**: Sensitivity of fragility curves (FC) and their 90% confidence intervals (CI FC) to (a) $h_0$ and (b) spatial distribution of buildings.

The impact of the density of the buildings is studied by re-running the 'BASE' scenario assuming that, in each town, the 10,000 buildings are located on the nodes of a wider 50×50 $m^2$ grid. By repeating the procedure outlined above, the mean and 90% confidence intervals of the empirical fragility curves constructed from this scenario are plotted against their 'BASE' counterparts in Fig. 4b. A wider grid means that the buildings cover a larger area. This leads to an IML range 11% larger than for the 'BASE'. In addition, the confidence intervals appear to be narrower, suggesting that surveying buildings far away from each other reduces the impact of the spatial correlation. For example, for PGA=0.20 g the width of the confidence limits is 20% smaller than for 'BASE'.

The sensitivity analyses so far used the entire building inventories in the five examined towns. In reality, surveying 50,000 buildings would be a time-consuming and expensive task, which leads to the questions:

can the accuracy of a large sample size be reached using fewer samples? And, if so, does the adopted sampling technique matter?

The impact of the sample size on empirical fragility curves is examined by considering two scenarios. The first scenario, termed 'N=5×100 (Ordered 1)', considers 100 buildings uniformly distributed on the 20×20 m²grid around the centre of each town, as presented in Fig. 5a. The second scenario, termed 'N=5×1000 (Ordered 1)', considers 1000 buildings uniformly distributed on the 20×20 m² grid around the centre of each town, as presented in Fig. 5b. For each building, the corresponding 1000 indicators generated for 'BASE' are assigned. For each realisation of indicators, empirical fragility curves are then constructed by fitting the probit model (Eq. (3)) to the 500 or 5000 indicators and their associated median PGA values obtained from GMPE1. Fig. 6 shows that the mean fragility curves for both scenarios are identical to their 'BASE' counterpart. However, the width of the confidence intervals varies considerably with the sample size. From Table 1, it can be inferred that the width of the 90% confidence intervals for 'N=5×100 (Ordered 1)' is 35% larger than for 'BASE'. This difference is reduced to 15% for 'N=5×1000 (Ordered 1)'.

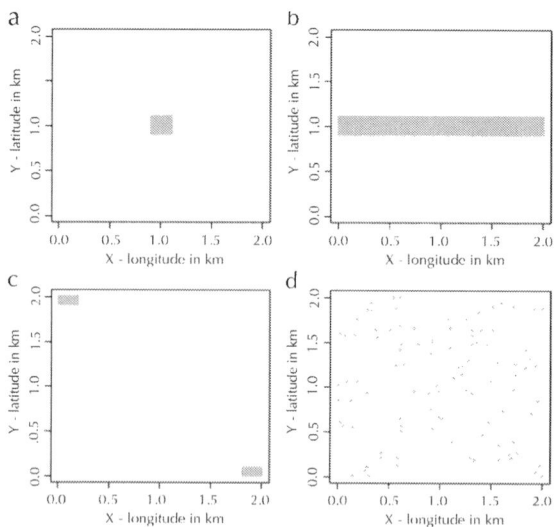

**Figure 5**: Distribution of 100 and 1000 selected buildings in town A for the four considered scenarios. (a) N=5x100(Order 1), (b) N=5x1000(Order 1), (c) N=5x100(Order 2), and (d) N=5x100(Random).

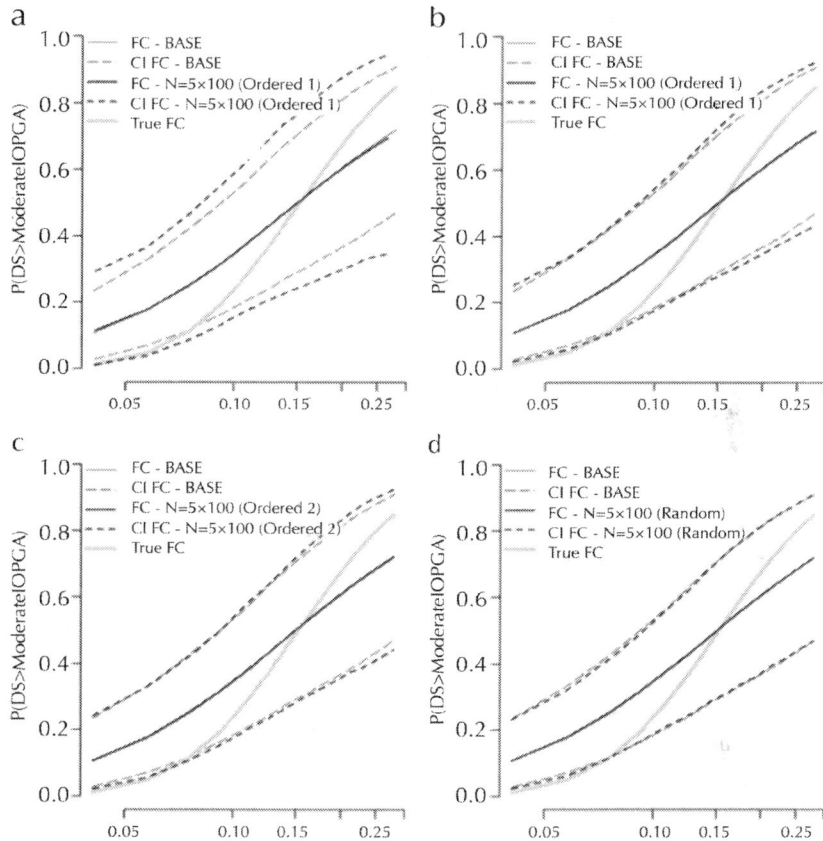

**Figure 6**: Sensitivity of fragility curves (FC) and their 90% confidence intervals (CI FC) to the number of buildings and their location aggregation. ((a)-(d), PGA in g).

The influence of the sampling technique adopted is examined next by examining whether the large uncertainty around the mean empirical fragility curve for scenario 'N=5×100 (Ordered 1)' is reduced by changing the sampling technique. This investigation is conducted by considering two further scenarios. The scenario, termed 'N=5×100 (Ordered 2)' assumes that 50 buildings uniformly distributed on the 20×20 m²grid are obtained from two different areas of each town, as presented in Fig. 5c. The scenario termed 'N=5×100 (Random)' randomly selects 100 buildings from each town (see Fig. 5d). Empirical fragility curves are constructed using the procedure adopted for the two aforementioned scenarios. Fig. 6 shows that the mean fragility curves for both scenarios are also identical to 'BASE'. In this case, the width of the confidence intervals appears to be closer to 'BASE'. From Table 1, it can be inferred that the width of the 90% confidence intervals for 'N=5×100 (Ordered 2)' and 'N=5×100 (Random)' for PGA=0.07 g is 13% and less than 10% the width of 'BASE', respectively. This suggests that the sampling technique adopted affects the results substantially. More importantly, relatively small carefully selected samples can yield results close to the curves obtained if every single affected building is surveyed.

Post-earthquake damage observations are often available in aggregated form. What is the effect of aggregation on the empirical fragility curves in the absence of strong-motion stations? This question is addressed by considering two scenarios with different levels of aggregation. The scenario termed 'aggregated=1 per town' aggregates the 10,000 buildings in each town to a single bin. The PGA for each of the five bins is estimated at the corresponding town's centre. This is a common assumption found in the literature. Similarly, the scenario termed 'aggregated=10×10 per town' aggregates the 10,000 buildings in each town to 100 equally sized sub-areas. The intensity measure level for each subarea is estimated at its centre. The mean and 90% confidence intervals of the empirical fragility curves for the two scenarios are presented in Fig. 7. The mean curve is identical for the two scenarios and they appear to be steeper than the 'BASE' mean curve, although the differences, especially at the lower end of the curves, remain significant. For example, for PGA=0.07 g the probability of exceedance for 'Aggregated=1 per town' is reduced, compared to its 'BASE' counterpart, to 92%. The confidence intervals appear to be significantly wider than for 'BASE'. Fig. 7b shows that the aggregated results are closer to the fully aggregated results, which is expected given that aggregation of buildings assumes that they are all subjected to the same ground motion. From Table 1, larger differences in the width of the confidence intervals can be observed for PGA=0.20 g. In particular, the 90% confidence intervals for 'Aggregated=1 per town' and 'Aggregated=10×10 per town' are 83% and 72% wider than for 'BASE'. This indicates that data aggregation leads to a significant loss of information, which appears to substantially increase the uncertainty in the empirical fragility curves. This observation seems to contradict the conclusions of Bal et al. [16] that the uncertainty introduced by the geographical resolution of observations of the mean repair-to-replacement ratio (MDR) is too small to justify a systematic effort to reduce it. It should be noted that Bal et al. [16] generated a large number of loss scenarios and they did not perform empirical fragility analysis. In particular, they generated random fields of ground motion intensity for a hypothetical town affected by a given earthquake and for each IML, the response of each building was simulated. A damage state was then assigned and finally the MDR was estimated and aggregated over the total number of buildings in the town. This aggregation might be the reason for the relatively small uncertainty around MDRs for each event. More research is needed in order to examine the effect of empirical fragility curves accounting for ground-motion uncertainty on the scenario loss.

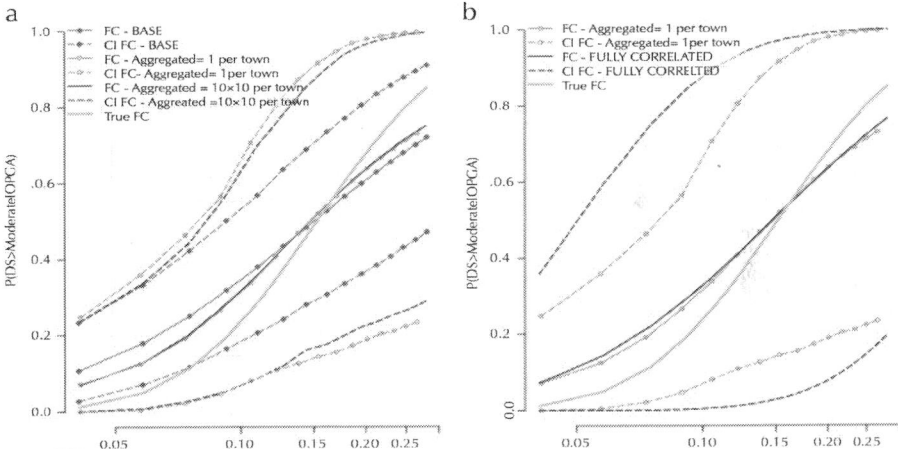

**Figure 7**: Sensitivity of fragility curves (FC) and their 90% confidence intervals (CI FC) to the level of aggregation ((a), and (b), PGA in g).

So far, the considerable influence of ground-motion variability on empirical fragility curves in the absence of ground motion recording stations has been highlighted. Nonetheless, the presence of ground motion recording stations (accelerographs) is expected to reduce this uncertainty. Does this mean that the presence of a relatively small number of stations can lead to more accurate empirical curves?

In order to address this question, the influence of the presence of a varying number of recording stations in each town is examined following the procedure described in Section 2.2.2. It should be noted that this procedure is computationally intensive. In order to improve its efficiency, the sensitivity analyses are based on the modification of the scenario 'N=5×100 (Random)'. In particular, each town has 100 randomly-distributed buildings. The damage in each building is characterised by the 1000 indicators obtained for the 'N=5×100 (Random)' case. For this analysis, however, the 'true' PGAs account for the presence of 1, 10, 50 and 100 stations per town located in some of the 100 selected buildings following the procedure outlined inSection 2.2.2. The procedure is repeated assuming that the correlation model required for kriging varies. In particular, the case where the spatial correlation is ignored is considered by setting $h_0$=0 km and the case where the 'true' intensity experienced by the 100 buildings is considered fully correlated is also simulated by assuming $h_0$=100,000 km.

The mean and 90% confidence intervals of the empirical fragility curves obtained assuming that the PGA values have been recorded (i.e. '5×100 st, COR.', '5×100 st, UNCOR.' or '5×100 st, FCOR.') in all 100 buildings for the three correlation models are presented in Fig. 8 together with their counterparts obtained for 'BASE' and the 'true' fragility curve. As expected, the results for the former three scenarios are identical. Their mean fragility curves are identical to the 'true' fragility curve corresponding to the moderate damage state. It should be noted that the confidence intervals for the three considered scenarios appear to be wider than their 'CHECK' counterparts due to the smaller number of buildings used (i.e. 500 instead of 50,000).

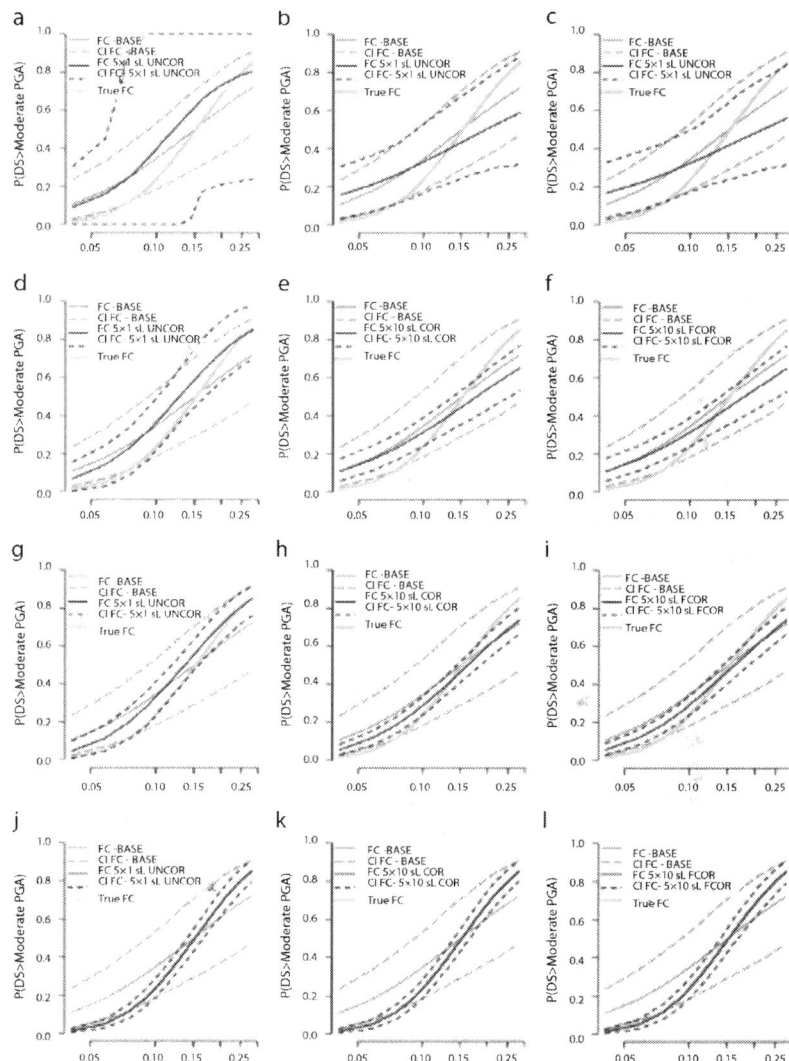

**Figure 8**: Fragility curves (FC) and their 90% confidence intervals (CI FC) corresponding to the moderate damage state assuming the presence of a varying number of ground-motion stations assuming three different correlation models ((a)-(l), PGA in g).

On the other hand, the presence of one station per town produces empirical fragility curves, whose mean and confidence intervals depend on the correlation model adopted. With regard to the mean fragility curves, the curve for scenario '5×1 st, UNCOR' appears to be closer to the 'true' fragility curve than that for '5×1 st, FCOR', Nonetheless, the differences between the mean curves and the 'true' remain significant in the lower tails for all cases. For example, for PGA=0.07 g, the probability of exceedance for '5×1 st, UNCOR' and '5×1 st, FCOR' is 129% and 174% higher than for its 'true' counterpart, respectively. The curve for scenario '5×1 st, COR' is included in the envelope formed by the two extreme correlation models and it is almost identical to its counterpart for the fully correlated scenario. The mean curve for 'BASE' is also included in this envelope. With regard to the confidence intervals, their width appears to be significantly wider for '5×1 st, UNCOR' and reduced for '5×1 st, FCOR'. In this case, the 90% confidence intervals for 'BASE' appear to be very close to their counterparts for '5×1 st, FCOR'. The aforementioned observations suggest that the presence of a very small number of stations, distributed in the affected area, does not improve the accuracy of the empirical fragility curves. To explore the reason behind this, the actual IMLs are plotted against their corresponding values estimated by kriging in Fig. 8. The closer the values to the 45° line, the better the prediction provided by a given number of stations in the area. From this figure, it can be seen that the presence of only one station per town leads to poor prediction of the actual IMLs and this affects the accuracy of the empirical fragility curves.

The presence of larger number of stations, i.e. '5×10 st, COR' and '5×50 st, COR', improves the prediction of the actual values (see Fig. 9) and this results in increasingly steeper mean fragility curves with increasingly narrower confidence intervals. In particular, for PGA=0.07 g, the width of the confidence intervals is 52% and 69%, for the two aforementioned scenarios, smaller than its 'BASE' counterpart. This indicates that the presence of ground-motion measurements at 10% or 50% of the surveyed buildings greatly improves the confidence around the mean empirical fragility curve. However, despite a decrease from 134% to 83% for the two aforementioned scenarios, the difference between the mean fragility curves and the 'true' fragility curve remains high. Similar observations can be noted for the other two correlation models. This lack of significant improvement in the mean empirical fragility curves provided by the extrapolation of PGAs from stations located in half of the total number of buildings (which is equivalent to 25 stations per km²) considered indicates that a very dense network of ground motion recording stations is required in order to construct reliable empirical fragility curves.

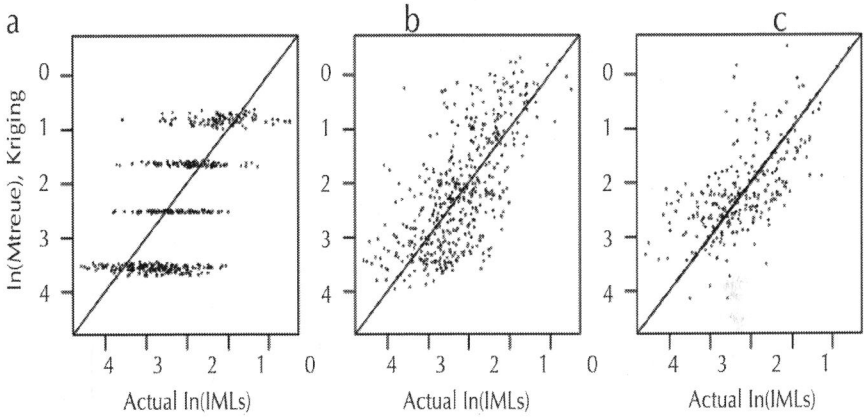

**Figure 9**: Actual ln(IMLs), for the scenario 'N=5×100 (random)', against the ln(IMLs) estimated by kriging (scenarios '5×1, 10 and 50st, COR').

# CONCLUSIONS

In this study, a series of experiments were conducted in order to assess the importance of modelling the variability and uncertainty in earthquake ground motions on empirical fragility curves. The following four main conclusions can be drawn from this study.

• The impact of variability in ground-motion levels is significant, and leads to considerably flatter fragility curves and wider confidence intervals. Further research is required in order to examine the impact of the uncertainty in the intensity measure levels in the seismic risk assessment's main products. These include, e.g., the estimation of the economic loss suffered by a group of buildings for a given event or the annual failure probability in a given location, estimated using the corresponding empirical fragility curves.

• There is a need for a very dense network of ground motion recording stations in order for the recorded data to reduce uncertainty in empirical fragility functions. This observation is in line with the main findings of Crowley et al. [9].

- The use of aggregated damage data, which is typical of existing empirical fragility curves, is found to increase substantially the uncertainty in the empirical fragility assessment. This raises questions about the accuracy of existing empirical fragility curves, but requires further research as only five towns and a single earthquake have been considered in this study.

- Finally, the sampling technique adopted in the collection of data can improve the accuracy of the empirical fragility curves. It was found that the use of a randomly selected, relatively small sample of buildings (e.g. 100 buildings per town) can potentially lead to improved fragility curves compared to using buildings from a single neighbourhood.

Overall, the findings of this study highlight, in line with similar appeals in the literature, the need for denser networks of ground motion recording stations in urban areas, without which the reliability of the empirical fragility curves is questionable. Low-cost Internet-enabled sensors, such as those in the Quake–Catcher Network [17], provide a possible cost-effective method of dramatically increasing the density of available ground-motion observations of damaging earthquakes. The results also highlight the important role that reconnaissance teams can play in the collection of small samples of high-resolution damage data for the construction of reliable empirical fragility curves.

# ACKNOWLEDGEMENTS

The authors would like to acknowledge the French Embassy in the London (UK) for supporting the Collaborative Science and Technology Workshop on "Seismic Fragility of Urban Buildings and Infrastructure" that led to this work. John Douglas's contributions to this study were funded by internal BRGM Projects, whereas those of Ioanna Ioannou and Tiziana Rossetto were funded by the EPSRC Project "Challenging RISK" (EP/K022377/1). We thank two anonymous reviewers for their detailed comments on a previous version of this paper.

# REFERENCES

1.  T. Rossetto, I. Ioannou, D.N. Grant, Existing empirical fragility and vulnerability relationships: compendium and guide for selectionGEM Foundation, Pavia, Italy (2013)

2.  D. Straub, Der A. Kiureghian, Improved seismic fragility modeling from empirical data, Struct Saf, 30 (2008), pp. 320–336

3.  T. Rossetto, I. Ioannou, D.N. Grant, Guidelines for empirical vulnerability assessmentGEM Foundation, Pavia, Italy (2014)

4.  YazcanU. Empirical vulnerability modeling considering geospatial ground motion variability. In: Proceedings of the 11th international conference on structural safety and reliability. New York, USA; 2013.

5.  Gehl, P, Douglas, J, Seyedi, D., Influence of the number of dynamic analyses on the accuracy of structural response estimates, Earthqu Spectra, 2014, http://dx.doi.org/10.1193/102912EQS320M.

6.  J. Douglas, Seismic network design to detect felt ground motions from induced seismicity, Soil Dyn Earthqu Eng, 48 (2013), pp. 193–197

7.  F.O. Strasser, J.J. Bommer, Review: strong ground motions—have we seen the worst? Bull Seismol Soc Am, 99 (2009), pp. 2613–2637

8.  P.J. Ribeiro Jr., P.J. Diggle, geoR: a package for geostatistical analysis, R-News, 1 (2001, 15–18)

9.  H. Crowley, P.J. Stafford, J. Bommer, Can earthquake loss models be validated using field observations? J Earthq Eng, 12 (2008), pp. 1078–1104

10. N. Jayaram, J.W. Baker, Correlation model for spatially distributed ground-motion intensities, Earthq Eng Struct Dyn, 38 (2009), pp. 1687–1708

11. M.A. Erberik, Generation of fragility curves for Turkish masonry buildings considering in-plane failure modes, Earthq Eng Struct Dyn, 37 (2008), pp. 387–405

12. S. Akkar, J.J. Bommer, Empirical equations for the prediction of PGA, PGV, and spectral accelerations in Europe, the Mediterranean Region, and the Middle East, Seismol Res Lett, 81 (2010), pp. 195–206

13.  C. Cauzzi, E. Faccioli, Broadband (0.05 to 20 s) prediction of displacement response spectra based on worldwide digital records, J Seismol, 12 (2008), pp. 453–475

14.  J.X. Zhao, J. Zhang, A. Asano, Y. Ohno, T. Oouchi, T. Takahashi, et al. Attenuation relations of strong ground motion in Japan using site classification based on predominant period, Bull Seismol Soc Am, 96 (2006), pp. 898–913

15.  E. Devalaud, F. Cotton, S. Akkar, F. Scherbaum, L. Danciu, C. Beauval, et al. Towards a ground-motion logic tree for probabilistic seismic hazard assessment in Europe, J Seismol, 16 (2012), pp. 451–473

16.  I.E. Bal, J.J. Bommer, P.J. Stafford, H. Crowley, R. Pinho, The influence of geographical resolution of urban exposure data in an earthquake loss model for Istanbul, Earthq Spectra, 26 (2010), pp. 619–634

17.  E.S. Cochran, J.F. Lawrence, C. Christensen, R. Jakka, The Quake–Catcher Network: citizen science expanding seismic horizons, Seismol Res Lett, 80 (2009), pp. 26–30

# Pile-Reinforcement Behavior of Cohesive Soil Slopes: Numerical Modeling and Centrifuge Testing

Liping Wang and Ga Zhang

State Key Laboratory of Hydroscience and Engineering, Tsinghua University, Beijing 100084, China

## ABSTRACT

Centrifuge model tests were conducted on pile-reinforced and unreinforced cohesive soil slopes to investigate the fundamental behavior and reinforcement mechanism. A finite element analysis model was established and confirmed to be effective in capturing the primary behavior of pile-reinforced slopes by comparing its predictions with experimental results. Thus, a comprehensive understanding of the stress-deformation response was obtained by combining the numerical and physical simulations. The response of pile-reinforced slope was indicated to be significantly affected by pile spacing, pile location, restriction style of pile end, and inclination of slope. The piles have a significant effect on the behavior of reinforced slope, and the influencing area was described using a continuous surface, denoted as W-surface. The reinforcement mechanism was described using two basic concepts, compression effect and shear effect, respectively, referring to the piles increasing the compression strain and decreasing the shear strain of the slope in comparison with the unreinforced slope. The pile-soil interaction induces significant compression effect in the inner zone near the piles; this effect is transferred to the upper part of the slope, with the shear effect becoming prominent to prevent possible sliding of unreinforced slope.

# INTRODUCTION

Landslides are one of the severest geologic hazards around the world, the prevention of which is of great interest to engineers and researchers. The stabilizing pile, an important reinforcement structure, has been widely used to support unstable slopes in the last few decades [1–3]. Many methods have been proposed to form a good basis for proper design of pile-reinforced slopes [4, 5]; however, there are a few important issues to be clarified for the application of such piles in slope engineering. Therefore, systematical investigations are required on the behavior and reinforcement mechanism of pile-reinforced slopes.

The behavior of pile-reinforced slope can be investigated by using a diverse range of research approaches, which can be generally divided into three categories: field observations, model tests, and numerical analyses. Field observation is an essential approach to obtain first-hand data of reinforced slopes. For example, long-term monitoring was used to analyze the bending moments and displacements of the piles that were employed for a railway embankment [2]. Nevertheless, boundary conditions or loading conditions cannot be easily changed in a field test, which restricts such an approach to the study of the reinforcement mechanism.

Model tests offer a powerful approach to investigate the behavior and failure mechanism of a reinforced slope by efficiently considering various factors. A series of 1 g model tests were used to investigate the behavior of a pile-stabilized sandy slope [6, 7]. The centrifuge model tests play an important role in such a category because they provide an accurate simulation of the gravity stress field and the gravity-related deformation process. Therefore, centrifuge modeling has been widely used to study reinforced slopes with different reinforcement structures, including piles, geomembranes, geotextiles, and soil nails [8–13]. The measurement was finite in the model tests due to the small size of the model and the limitation of measurement technology; for example, the stress state of the slope cannot be measured with sufficient accuracy.

Numerical analysis can yield comprehensive information about the response of the slope; thus, a few different types of numerical methods were developed or used to study the reinforced slopes, such as the limit equilibrium method [14, 15], limit analysis [16], the finite-element method [17], and other rigorous or simplified methods [18]. For example, Won et al. compared the predictions by limit equilibrium analysis and three-dimensional numerical analysis involving a shear strength reduction technique for a slope-pile system [19]. A three-dimensional numerical analysis was used to investigate the influence of sleeving on the pile performance in a sloping ground [20, 21]. The FLAC3D program was used to analyze the response of piles in an embankment slope with a translational failure mode, and the results showed that the pile-soil relative stiffness has a significant influence upon the piles' failure mode [22]. Except for the algorithm, the effectiveness of numerical analysis of reinforced slope is also significantly affected by several aspects, such as the soil model, the soil-structure interface model, and their parameters, which should be acknowledged not to be sufficiently reliable due to the complexity of this problem.

Most previous investigations of the pile-reinforced slope focused on the response of the slope and piles as well as on the influence factors; from those, a few useful conclusions have been achieved. However, the response of the pile-reinforced slope under various types of load applications has not been illustrated in a fully comprehensive view. Moreover, the stabilizing mechanism—for example, why the avoidance of failure is induced by the piles—has not yet been adequately discovered. In other words, further study is needed to clarify how the local pile-soil interaction affects the deformation field of the entire slope and, therefore, increases the stability level. In addition, the deformation trends and main influence factors of the behavior of pile-reinforced slopes have not been fully discovered.

Based on the understanding of the main features of the numerical methods and model tests, an effective approach may be to combine both of the above methods to acquire a comprehensive description of the pile-reinforced slopes. Therefore, the observations from the model tests can be supplemented by a numerical method that has been verified by the model tests. The objective of this paper is to conduct such an attempt for the purpose of reinforcement mechanism analysis of pile-reinforced slope, including (1) to conduct centrifuge model tests of a cohesive soil slope using stabilizing piles, in comparison with the unreinforced slope; (2) to present a numerical scheme of the pile-reinforced slope and to confirm its effectiveness by simulating the centrifuge tests; (3) to analyze the behavior of the reinforced slope by using the test observations and numerical analysis; (4) to describe the reinforcement mechanism; and (5) to discuss the main factors that influence the behavior of reinforced slopes.

# CENTRIFUGE MODEL TESTS

The centrifuge model tests were conducted using the 50 g ton geotechnical centrifuge at Tsinghua University.

## Model

The soil of the slope model was retrieved directly from the soil mountain of the Beijing Olympic Forest Park. The average grain size of the soil is 0.03 mm, and the plastic limit and liquid limit are 5% and 18%, respectively.

Figure 1 shows the photographic and schematic views of the model slope, reinforced using stabilizing piles. The unreinforced slope is identical except for removal of the piles. The model container for the tests, made of aluminum alloy, is 50 cm long, 20 cm wide, and 35 cm high. A transparent Lucite window was installed in one container side to observe the deformation process of the soil.

(a)Photograph

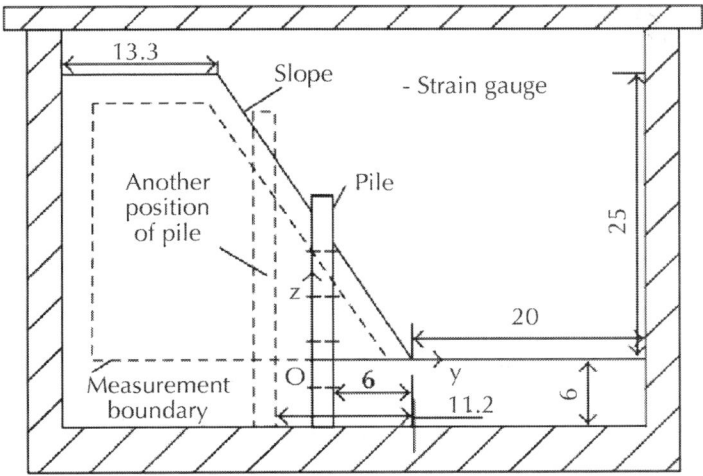

(b) Elevation view

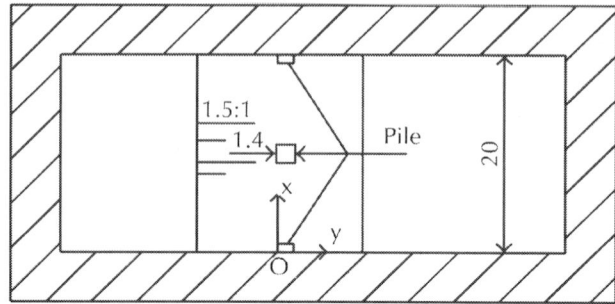

(c) Vertical view

**Figure 1:** Schematic view of the model of reinforced slope (unit: cm).

The soil was compacted into the container by a 6-cm-thickness layer, with a dry density and water content of $1.4\,g/cm^3$ and 16%, respectively. The slope was obtained by removing the redundant soil, with an inclination of 1.5:1 and height of 25 cm. A 6 cm high horizontal soil layer under the slope was set to diminish the influence of the bottom container. In addition, silicone oil was painted on both sides of the container to decrease the friction on the slope.

A hollow square pipe, made of steel with an elastic modulus of 210 GPa, was used to simulate the stabilizing pile of the reinforced slope. The pipe was 1.4 cm along the side length of the section, with a wall thickness of 1.5 mm. This is equivalent to a prototype pile with a side length of 0.7 m at 50g-level. These piles were inserted in the slope, without special fixation, at a single row 10 cm apart and 6 cm far away from the slope toe. Half-section piles were used near both container sides to approximate the plane-strain condition (Figure 1(c)).

# Measurements

An image-record and displacement measurement system was used to record the images of soil during the centrifuge tests [23], which are used to determine the displacement history of an arbitrary point of soil, without disturbing the soil itself, by an image-correlation-analysis algorithm [24]. The effectiveness of this measurement system was realized by embedding white particles laterally in the soil (Figure 1(a)). The measurement accuracy can reach 0.02 mm based on the model dimension for the centrifuge tests in this paper. In addition, a few patterns with significant grey difference were affixed onto the pile to obtain displacement history of the pile on the basis of image analysis. The area within the dotted line was used for displacement measurements owing to the requirement of the measurement system (Figure 1(b)); the main deformation zone can be covered. Cartesian coordinates were established with the origin as the intersection between the slope bottom and inner sidewall of the pile; positive directions were specified (Figures 1(b) and 1(c)).

Four pairs of strain gauges were attached to the inner walls of the middle pile to measure the strain distribution along the shaft (Figure 1(b)). They can be used to derive the bending moment and axial load of the pile.

## Test Procedure

The model slope was installed on the centrifuge machine, and increasing centrifugal acceleration was applied at steps of 5 g. Each acceleration step was maintained for several minutes until the deformation of the slope became stable. This process was terminated at 50 g-level when the unreinforced slope exhibited significant failure.

## Observations

It should be noted that all of the results are based on the model dimension in this paper. Figure 2 shows the horizontal displacement distribution of the pile-reinforced and unreinforced slopes at 50 g-level; the borders in the figures were designated as the dotted area in Figure 1(b), but these do not correspond to the actual slope borders. It can be seen that a significant landslide occurred in the unreinforced slope when the centrifuge acceleration reached 50 g (Figure 2(a)): there was significant concentration of deformation so that the slide body can be easily distinguished from the base body via the contour lines. On the other hand, the reinforced slope only exhibited significant deformation due to the increase of centrifugal acceleration; a landslide was avoided at 50 g-level (Figure 2(b)). This indicated that the piles significantly increased the stability level of the slope. Moreover, the deformation of the reinforced slope was significantly smaller than that of the unreinforced slope, demonstrating that the piles had a significant effect on the deformation of the slope.

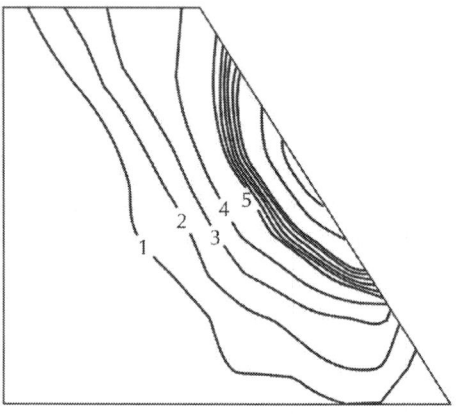

a. Unreinforced slope at failure state

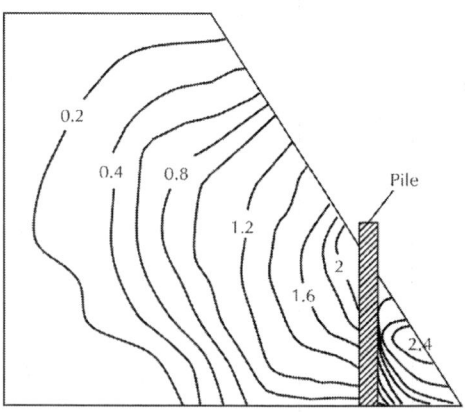

b. Pile-reinforced slope

**Figure 2:** Horizontal displacement contours of slopes at 50 g-level (unit: mm).

# NUMERICAL ANALYSIS

## Analysis Model

A three-dimensional finite element method was used to simulate the pile-reinforced slope under the condition of the centrifuge model test based on geotechnical stress-strain FEM software, termed TOSS3D that has been widely used for embankments in China [25]. A new iterative routine was developed to simulate the successive increase of centrifugal acceleration in the software. The explicit increment scheme was used in the nonlinear static FE analysis. A substep was divided into several subincrements to simulate the nonlinear loading. An iterative algorithm was employed to obtain the stress-strain states of the slope within a subincrement, with a trial algorithm used to judge loading states of the geomaterials and contact states of the interface.

A three-dimensional FE mesh was established with accurate simulation of the slope model for centrifuge tests (Figure 3). The soil was described using hexahedron elements with eight nodes. The soil was described using an elastoplasticity model that can reasonably capture the dilatancy behavior of the soil [25]. The model parameters were determined from triaxial compression tests and adjusted slightly in the numerical analysis for a better fit to the test observations. The parameters and their values are listed in Table 1 and their definitions could be referred to [25]. It should be noted that the cohesion strength parameter, $c$, was a bit greater than the empirical. This may be partially because the boundary effect on the model slope in the centrifuge model tests was considered by the strength parameters of the soil. The interface elements were set between the pile and neighboring soil and between the container sides and neighboring soil. The interface was described using an elastoplasticity damage model, which provides a unified description of monotonic and cyclic behavior, including volumetric behavior [26]. This model was used for many soil-structure systems, such as high concrete-faced rockfill dams [27]. The model parameters were determined by a series of shear tests under constant normal stress conditions. The parameters and their values are listed in Table 2 and their definitions could be referred to [26]. The pile was described using a linear elastic model, with elastic modulus of 210 GPa and Poisson's ratio of 0.3. A soft element set on the pile end to realize the movement of pile. Moreover, another case, with the pile located in the upper slope, as shown in Figure 1(b), was also considered in the mesh. The boundaries of the model slope were all fixed (Figure 3). The mesh was finally obtained according to the symmetry of this problem, involving 8448 nodes and 7140 elements in total.

**Table 1:** Model parameters of the elastoplasticity model for soil

| C(kPa) | $(\phi_0 °)$ | $R_f$ | K | n | G | F | D |
|---|---|---|---|---|---|---|---|
| 50 | 35 | 0.7 | 92 | 0.15 | 0.05 | 0.15 | 1.5 |

**Table 2:** EPDI model parameters of the interface

| 12 | $\alpha$ | $\beta$ | $\phi_0(°)$ | $G_0$ | $n_0$ | $\mu_0$ | $M_0$ | $K_0$ | $^M k_0$ | $C_e$ | $C_0$ |
|---|---|---|---|---|---|---|---|---|---|---|---|
| 0.15 | 500 | 1.0 | 33 | 30 | 0.5 | 300 | 0.3 | 0 | 0.1 | 0.005 | 0.005 |

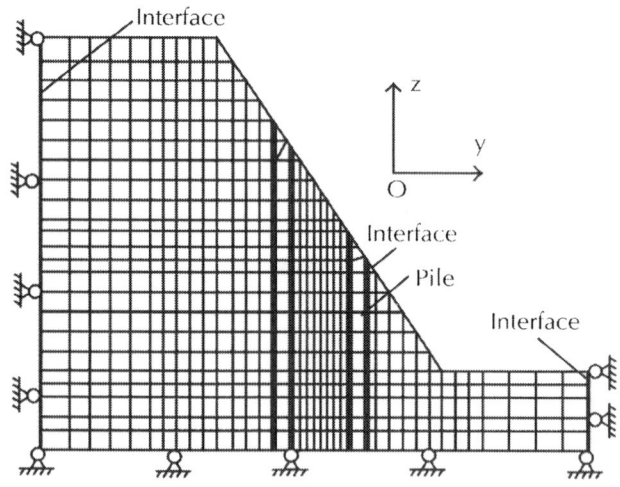

(a) Elevation view

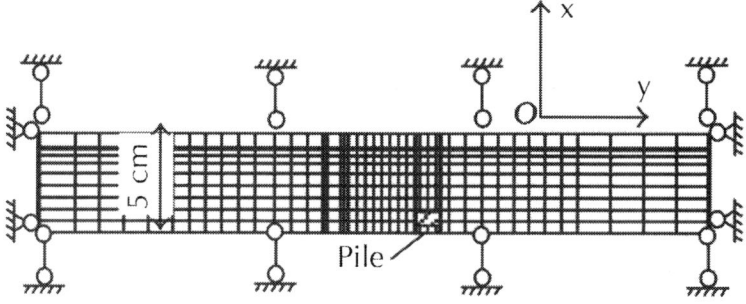

(b) Vertical view

**Figure 3:** Mesh and boundary of the slope for numerical analysis.

# Verification

The numerical predictions of displacement response of the reinforced slope were compared with the measurements of the centrifuge model test to verify the effectiveness of the numerical analysis.

Figure 4 shows a comparison of test results and numerical predictions of the contours of displacement of the reinforced slope at 50 g-level. It can be seen that the predicted curve showed a good fit to the test result; this demonstrates that the numerical method provides a reasonable description of the overall performance of the slope. Close comparisons of displacement distribution between the test results and numerical predictions were made on several vertical sections (Figure 5). The horizontal and vertical displacements exhibited the maximum at the middle and top of the slope, respectively. These comparisons showed that the numerical prediction curves were in satisfactory agreement with the test results at different locations. In addition, the vertical displacement histories of a typical point of the slope indicated that the vertical displacement increased with increasing centrifugal acceleration (Figure 6), and the numerical prediction showed a good fit to the test observation.

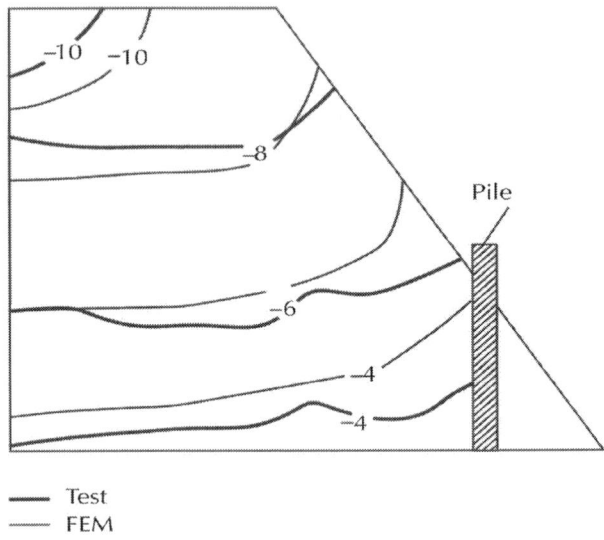

(a) Vertical displacement

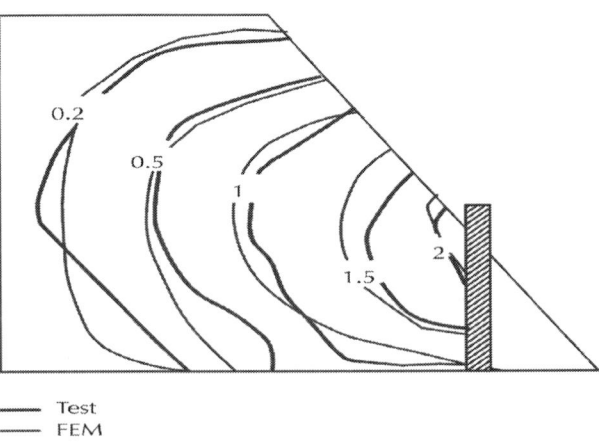

(b) Horizontal displacemen

**Figure 4:** Displacement distribution comparison of numerical analysis and test results (unit: mm).

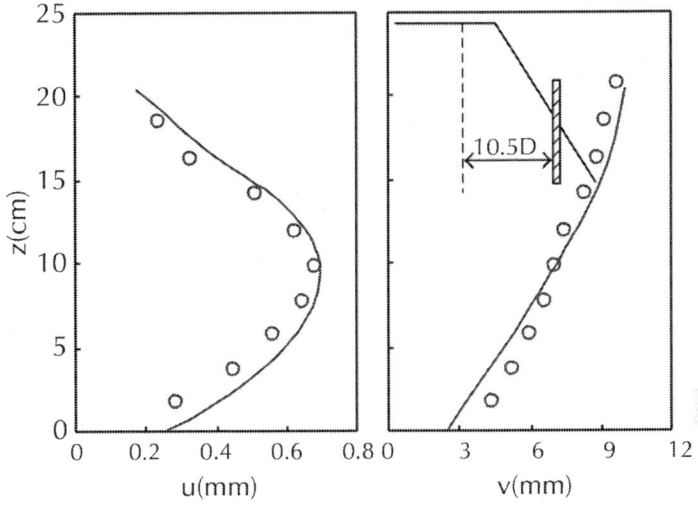

a. y=-10.5D

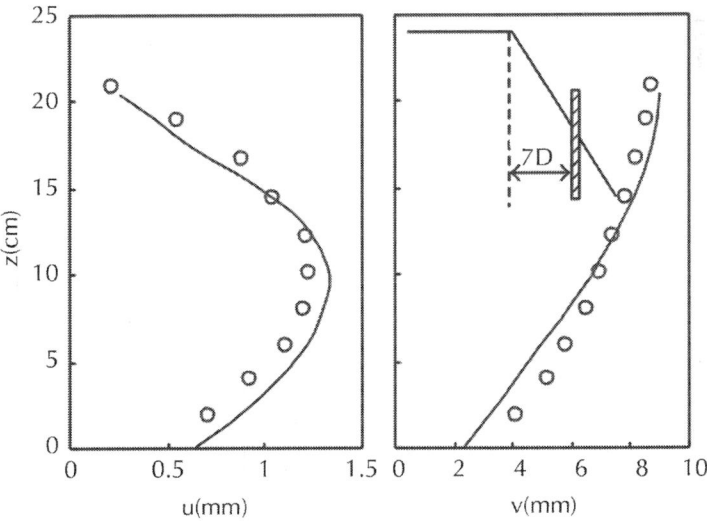

b. y=-7D passing by slope shoulder

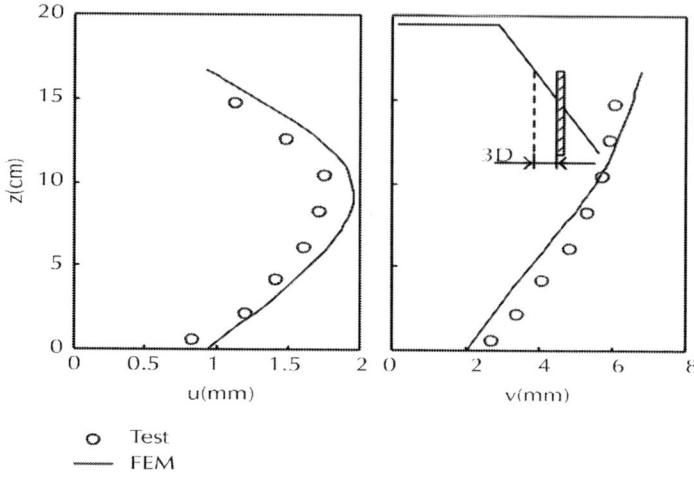

c.  y =-3D

**Figure 5:** Displacement distribution comparison of numerical analysis and test results on different vertical sections. u: horizontal displacement; v vertical displacement; D: pile diameter.

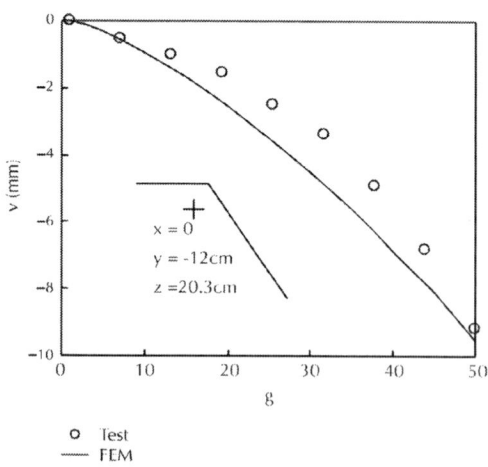

**Figure 6:** Vertical displacement history of a typical point by numerical analysis and test results. v: vertical displacement; g: g-level.

The response of the pile was also used for the comparison of numerical analysis and test results (Figure 7); a satisfactory fit can be found. It should be noted that we used the difference between the vertical strains obtained from the strain gauges on the left and right sides of the pile to consider the bending behavior (Figure 7(b)); this difference exhibited the maximum at the middle part.

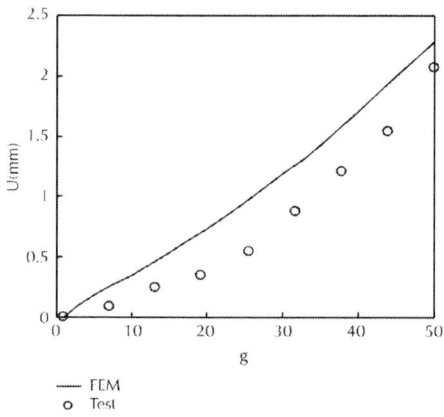

a. Horizontal displacement of pile top

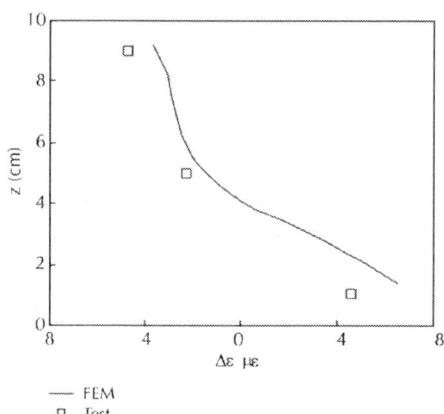

b. Vertical strain distribution

**Figure 7:** Response of the pile by numerical analysis and test results. u: horizontal displacement of pile top; $\Delta_\varepsilon$: vertical strain difference between right and left sides of pile.

According to the comparison results, it can be concluded that the numerical analysis is effective in capturing the primary behavior of a pile-reinforced slope. The numerical results can be further used to analyze the stress response of the reinforced slope, which is important for understanding the reinforcement behavior but difficult to be measured in the centrifuge model tests. Thus, a comprehensive stress-deformation response can be obtained by combining the numerical and physical simulations.

# STRESS-DEFORMATION BEHAVIOR

The numerical results have shown that the stress-displacement response of the reinforced slope is approximated for different sections, whether the piles pass through or not if the pile space is not large, as for most practical cases. Therefore, the behavior of the pile-reinforced slope was analyzed based mainly on the stress-displacement distribution of the lateral side of the slope that the pile passed through, so that the measurement results of the centrifuge model test can be combined with the predictions of numerical analysis.

The stress distribution of the slope was illustrated using numerical analysis at 50 g-level (Figure 8); there was significant stress concentration near the piles. This demonstrated that the piles had a significant effect on the stress state of the neighboring slope, as did the displacement measured by the tests (Figure 2). Figure 9 shows the stress histories of a typical point on the slope. The magnitude of stress increased with increasing centrifugal acceleration; similar rules can be found in the displacement histories (Figure 6). It can be concluded that the stress-deformation response at 50 g-level can be used as a representative time for further analysis.

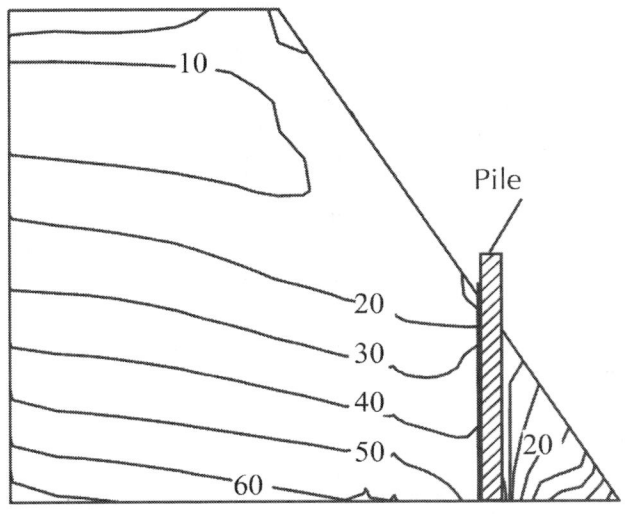

(a) Horizontal stress

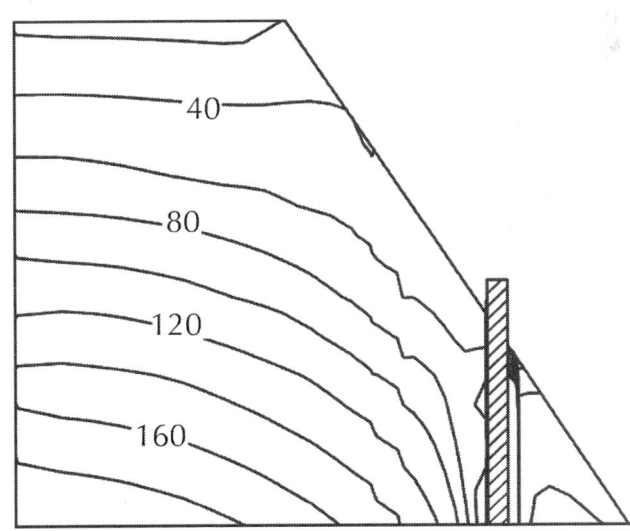

b. Vertical stress

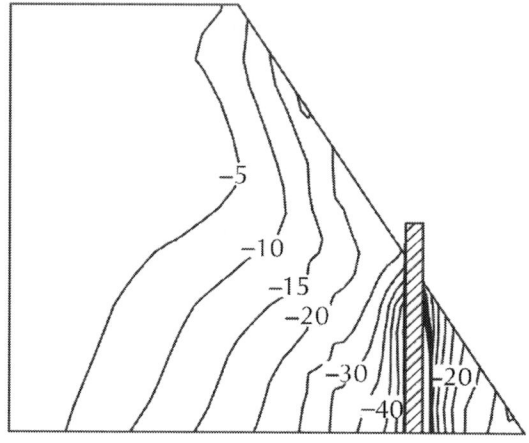

c. Shear stress

**Figure 8:** Stress distribution of the slope at 50 g-level by numerical analysis (unit: kPa).

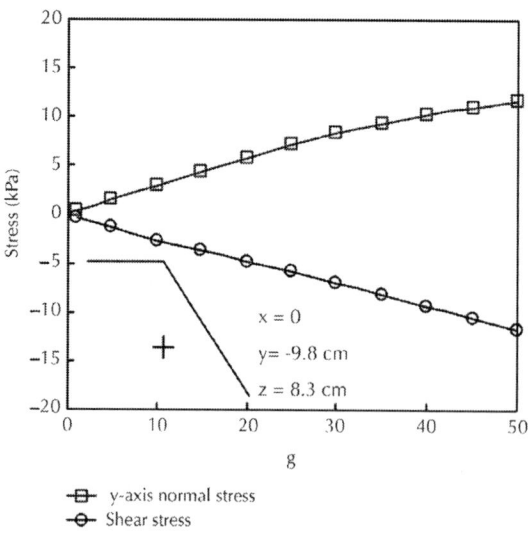

**Figure 9:** Stress history of a typical point by numerical analysis. G: g-level.

# Significant Influence Surface

The piles have a more significant influence on the horizontal displacement of the slope than on the vertical displacement, according to the comparison of reinforced and unreinforced slopes. Thus, the distributions of horizontal displacement at horizontal lines of five altitudes were carefully analyzed by comparing the reinforced and unreinforced slopes, covering the overall slope from top to bottom (Figure 10).

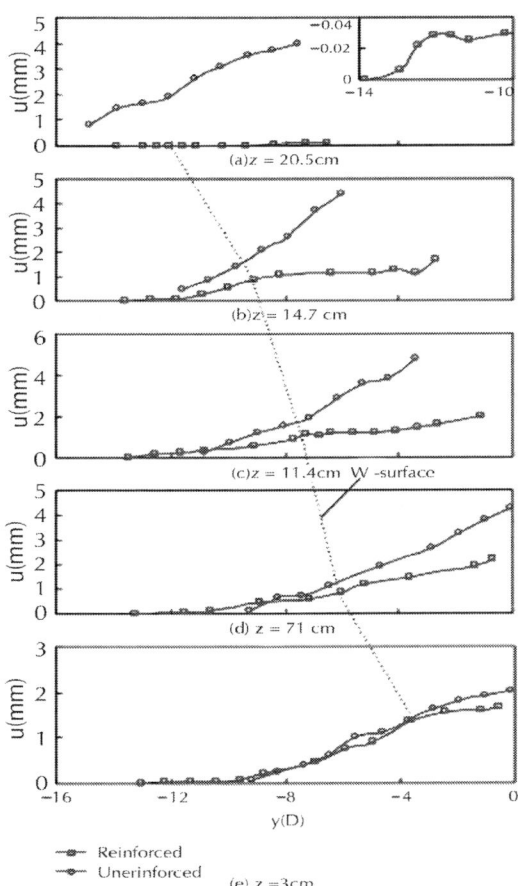

**Figure 10:** Horizontal displacement distributions on horizontal sections at 50 g-level by test observation. u: horizontal displacement; D: pile diameter.

A close examination of displacement distribution at a horizontal line, z = 7.1 cm, showed a significant difference between the reinforced and unreinforced slopes (Figure 10(d)). For the pile-reinforced slope, an evident inflection occurred near the pile. On the left side of the inflection, the horizontal displacement increased significantly from the inner slope area to the piles, whereas this rate of increase became relatively small on the right side. On the other hand, for the unreinforced slope, the horizontal displacement increased from the inner slope at an approximately constant rate near the piles. It can be concluded that the piles significantly changed the displacement distribution of the slope at a certain area near the piles, and this inflection can be regarded as a boundary point to indicate that the piles significantly affected the deformation.

The inflections can be found in the displacement distribution curves of the reinforced slope at all altitudes (Figure 10), including the area above the piles (Figures 10(a)–10(c)). Therefore, a continuous surface was obtained by connecting these inflections using a curve, as shown in Figure 10 by the dotted line, denoted as the W-surface in this paper. The horizontal displacement of the pile-reinforced slope exhibited different distribution rules on different sides of the W-surface.

Similar to the horizontal displacement, the horizontal stress can also be used to reflect the effect of the piles on the slope. Figure 11 shows the horizontal stress distributions on horizontal lines at two typical altitudes, located near and above the piles, with the position of the W-surface being determined from the displacement distribution. It can be seen that the stress distribution curve exhibited a significant change near the W-surface. This demonstrated that the W-surface outlines the area where the piles have a significant effect on the slope, including the displacement and stress response. Thus, the W-surface can be used as an important index to describe the influence of piles on stress-deformation behavior of the reinforced slope.

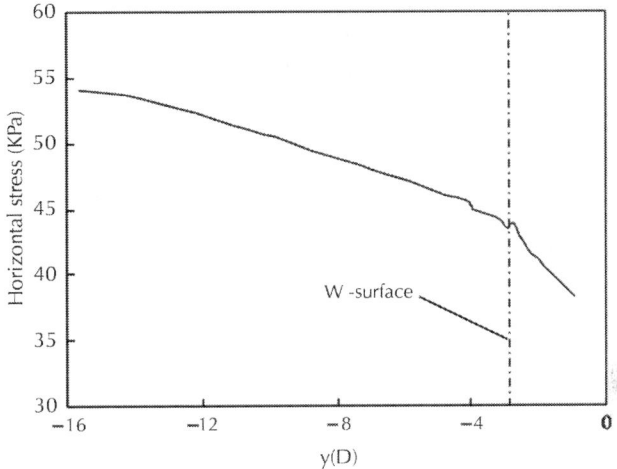

a z=3.5 cm

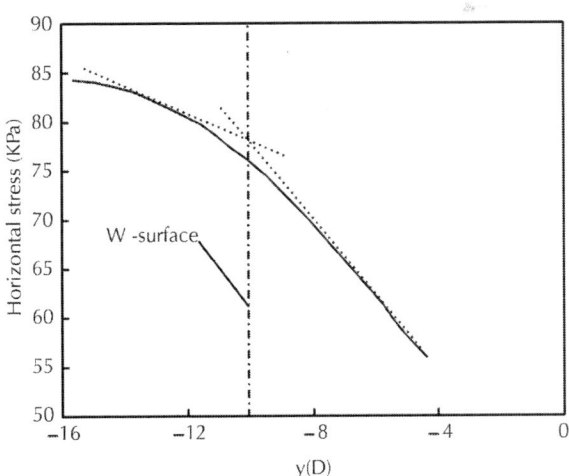

b. z=16.5cm

**Figure 11:** Stress distribution on a horizontal line by numerical analysis.

## Division of Zones

According to the W-surface, the reinforced slope behind the piles can be divided into two zones: one is where the slope is significantly affected by the piles, while the other is where the slope is insignificantly affected. The former was further examined by using the horizontal stress distribution of the slope at different vertical sections (Figure 12). The horizontal stress decreased with increasing altitude from the slope bottom; however, there was a significant inflection in the upper part of the slope. From this inflection, the change trend of the stress noticeably altered. It can be concluded that the stress states of the slope were significantly different on the different sides of the inflection, and the reason can be attributed to the effect of the piles.

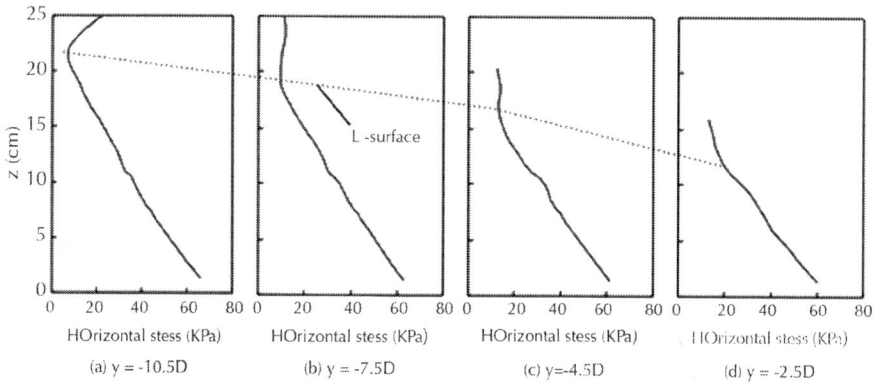

**Figure 12:** Stress distributions on different vertical sections at 50g-level by numerical analysis.

Thus, another continuous surface was yielded by connecting these inflections using a curve, as the dotted line showed in Figure 12. This surface, denoted as the L-surface in this paper, indicates the boundary where the effect of the piles exhibited different features on different sides. Accordingly, the slope can be divided into three zones behind the piles according to these two surfaces (Figure 13).

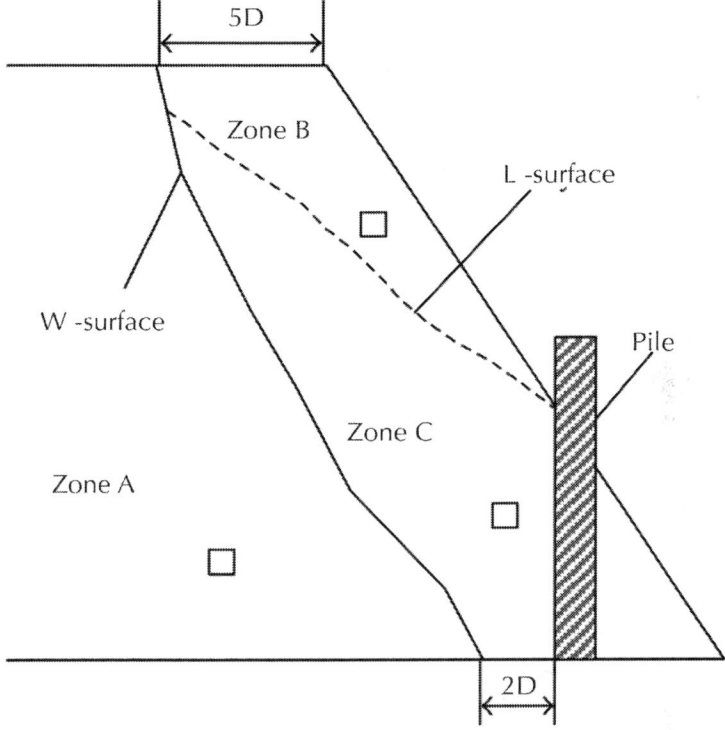

**Figure 13:** Division of zones with different features. D: pile diameter.

The stress-deformation behavior exhibited different features in different zones. Zone A is the area at the left side of the W-surface, far from the piles. Thus, the piles had a small effect on this zone. For example, the horizontal displacement of the reinforced slope showed a similar distribution to the unreinforced slope in zone A. Zone B is above the L-surface and has a free surface. The change of deformation of the slope became flat in this zone; this demonstrated that the piles significantly restricted the deformation. Zone C is at the right side of the W-surface and directly contacts the piles. The piles significantly reduced the horizontal displacement of the soil in this zone; thus, the displacement distribution showed a flat curve. Moreover, the horizontal stress exhibited a significant change with increasing distance from the piles. This demonstrated that the piles induced significant compression effect in this zone.

# REINFORCEMENT MECHANISM

The strain of a soil element within the slope was introduced to analyze the pile-reinforcement mechanism of the slope by comparing the reinforced and unreinforced slopes. This strain can be determined using the measured displacement of the centrifuge model tests because the measurement may be more reliable than the numerical result, especially at the failure state of the unreinforced slope. A two-dimensional, four-node square isoparametric element, 1 cm long, was used for strain analysis. The strain at the center was thought to be the one of this element that was assumed to be uniform within the element.

The deformation of soil can be divided into two independent components owing to shear and compression applications. This implies that shear and compression play different roles in the deformation and failure of a slope. For example, the formation of a slip surface may significantly depend on the increase in shear strain, whereas a tensional crack, which commonly occurs at the top of a slope, is induced mainly by a decrease in compression strain. It can be concluded that the piles significantly changed the strain state of the slope and thus increased the stability level of the slope. Therefore, the pile-reinforcement mechanism can be analyzed using the effects of the piles on the shear and compression strains.

The deformation due to shear was indicated by the slope-direction shear strain. The horizontal compression strain was used to indicate the deformation due to compression. Figure 14 shows the strain histories of typical elements in the three zones (positions are shown in Figure 13); the elements of the unreinforced slope, corresponding to those in the reinforced slope, are also presented for comparison. It should be noted that the horizontal compression strains of these elements were all negative, indicating dilation in the horizontal direction.

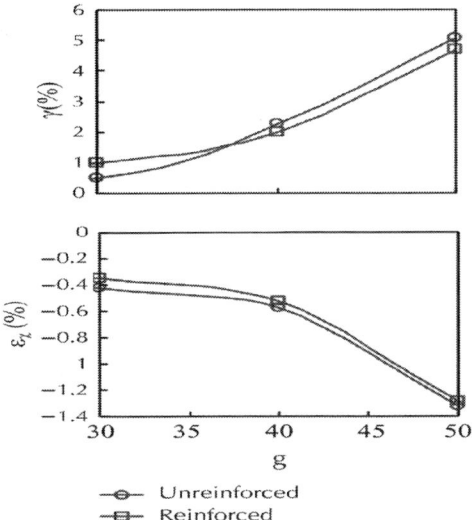

(a) Zone A

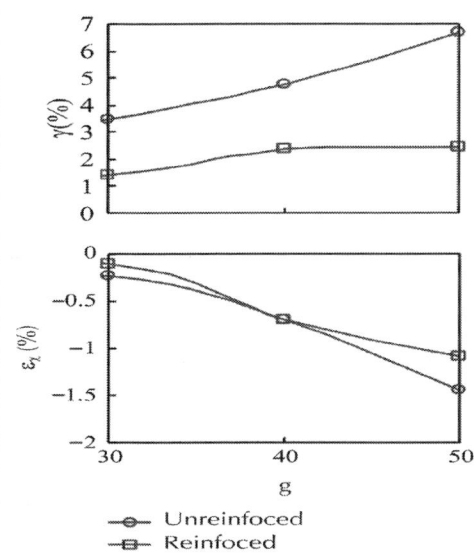

b) Zone B

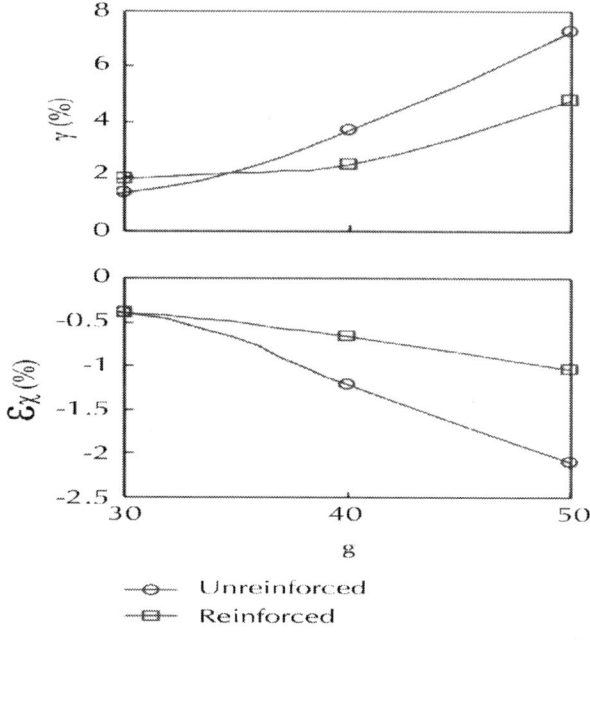

(c) Zone C

**Figure 14:** Strain history of typical elements of different zones by test observation. $\gamma$: slope-direction shear strain; $\varepsilon_x$: horizontal compression strain; g: g-level.

In zone B, the shear strain of a typical element in the unreinforced slope increased rapidly with increasing centrifugal acceleration (Figure 14(b)). This strain reached a significant magnitude at 50 g-level when the landslide, just across this element, occurred. However, the shear strain increased at a smaller rate and reached a lower level far from failure if the piles were used. Thus, a basic concept, shear effect, was introduced to describe that shear strain was decreased due to the effect of the piles. In other words, a significant shear effect of piles in zone B arrested the formation of a slip surface. Accordingly, the horizontal compression strain increased due to the piles, which was described using another basic concept, compression effect. There was also significant compression effect of piles in zone B; however, it can be concluded that the shear effect was primary.

The history of the compression strain showed that this strain in the reinforced slope was significantly larger than that of the unreinforced one; this demonstrated that there was significant compression effect of piles in zone C (Figure 14(c)). Closer examination showed that the compression effect was more significant than the shear effect, though the shear effect of the piles was also distinct.

The compression and shear strains both exhibited minor differences in zone A for the reinforced and unreinforced slopes (Figure 14(a)); this indicated that both the shear effect and compression effect were negligible in this zone.

Figure 15 shows the earth pressure on the inner side of the pile, obtained by numerical analysis. The earth pressure decreased with increasing altitude, similar to a distribution of earth pressure for a retaining wall. Combined with the strain analysis results at the overall slope level, the reinforcement mechanism can be demonstrated using the shear effect and compression effect of piles, which were of different levels in different zones, as follows: the pile and neighboring soil exhibited a significant interaction due to loading. This interaction was transferred to the adjacent zone and induced a significant compression effect (zone C). This compression effect transferred upwards and caused a prominent shear effect (zone B), arresting the possible sliding and increasing the stability level of the slope.

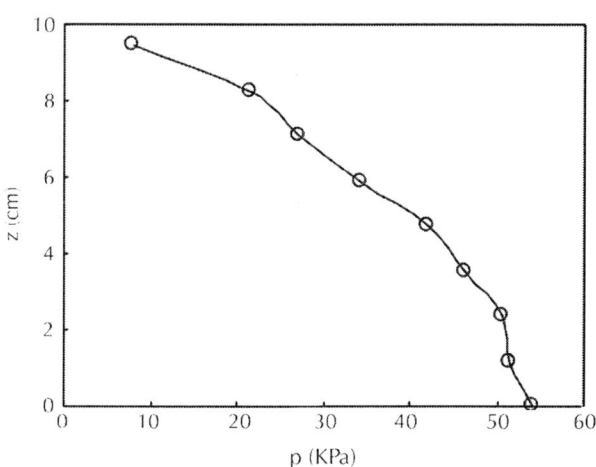

**Figure 15:** Earth pressure on the pile by numerical analysis. p: earth pressure.

# INFLUENCE FACTORS

The numerical analysis was used to compute different cases to discuss the influence factors of the behavior of pile-reinforced slopes by altering several factors based on the centrifuge test condition, including pile spacing, pile location, restriction style of pile end, and inclination of slope. According to previous analysis, the horizontal displacement of the lateral side of the slope that the piles pass across is used as the characteristic index to illustrate the effect of these factors.

## Pile Spacing

The pile spacing was reduced to 3D (pile diameter) from approximately 7D of the test condition. Figure 16 compares the horizontal displacements according to the different pile spacings, both obtained from numerical analysis, to discuss the influence rules. The horizontal displacement of the slope exhibited a small decrease when the pile spacing was reduced; such a difference decreased with increasing distance from the piles. The fundamental rules of horizontal displacement were consistent for different pile spacings.

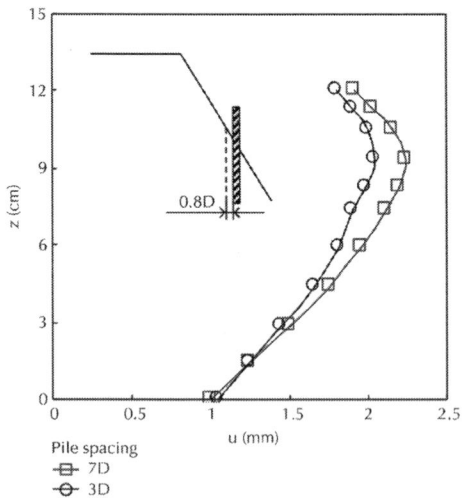

a) y=-0.8D

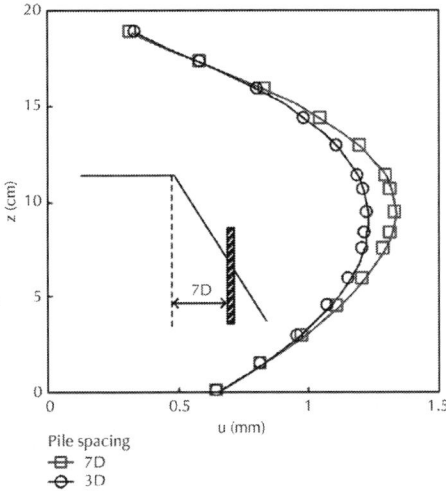

b) y=-7Dpassing by slope shoulder

**Figure 16:** Influence of pile spacing by numerical analysis. D: pile diameter; u: horizontal displacement.

# Location of Pile

The piles were moved upwards several centimeters in the comparison case, as described using the dashed line in Figure 1(b). The horizontal displacements according to new and original locations of piles were obtained from numerical analysis (Figure 17). The horizontal displacement of the slope exhibited a significant decrease near the piles if the pile was located at an upper position, which may be partly attributed to the new location of piles being farther from the free surface of the slope. The difference decreased with increasing distance from the slope surface and can be ignored at the section passing by the slope shoulder (Figure 17(c)). It can be derived that the distributions of horizontal displacement of the pile-reinforced slope were consistent for different pile locations.

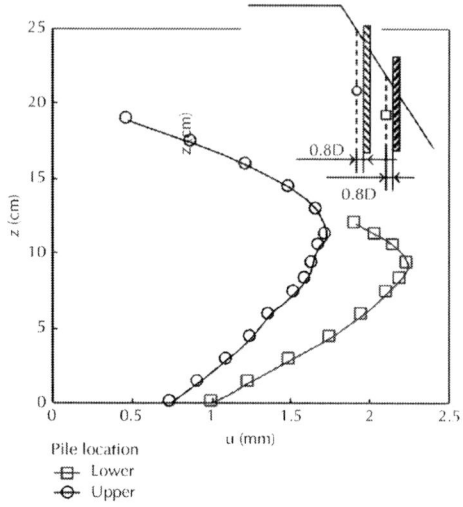

(a) 0.8D from pile

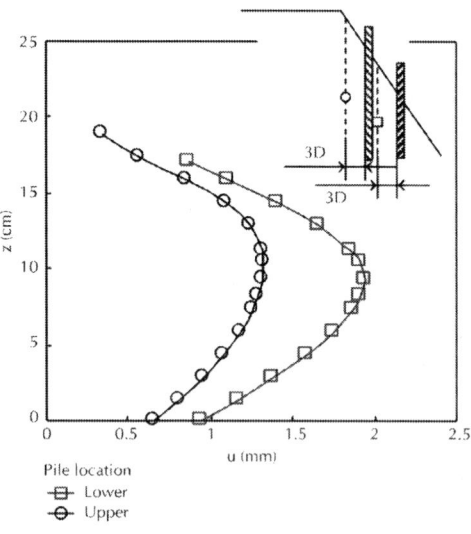

(b) 3D from pile

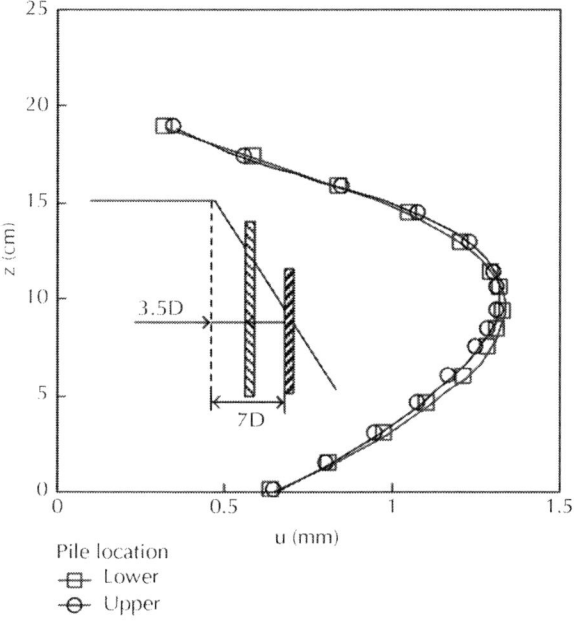

Pile location
 ⊟ Lower
 ⊖ Upper

a) y=-7Dpassing by slope shoulder

**Figure 17:** Influence of positions of the pile by numerical analysis. D: pile diameter; u: horizontal displacement.

## Restriction Style of Pile End

The pile ends were all fixed on the container to prevent the relative movement between the piles and the container bottom, different from the test condition that the piles may move along the container bottom. Figure 18 compares the horizontal displacements according to the different restriction styles, both obtained from numerical analysis. The horizontal displacement of the slope significantly decreased, especially near the piles, when the pile ends were fixed. Overall, the fundamental rules of horizontal displacement of pile-reinforced slope can be found to be consistent regardless whether the restriction style of pile end was altered.

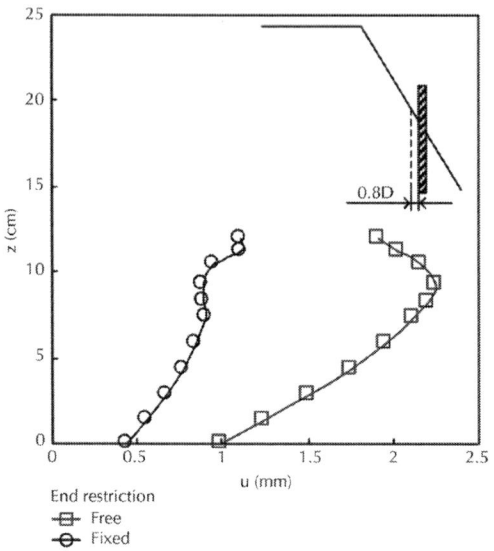

a) y=-0.8D

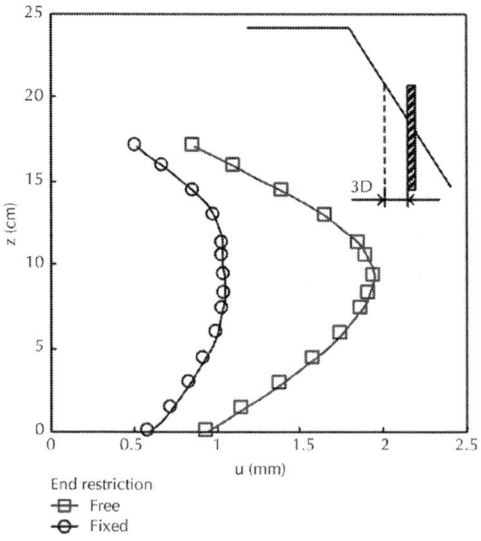

(b)y=-3D

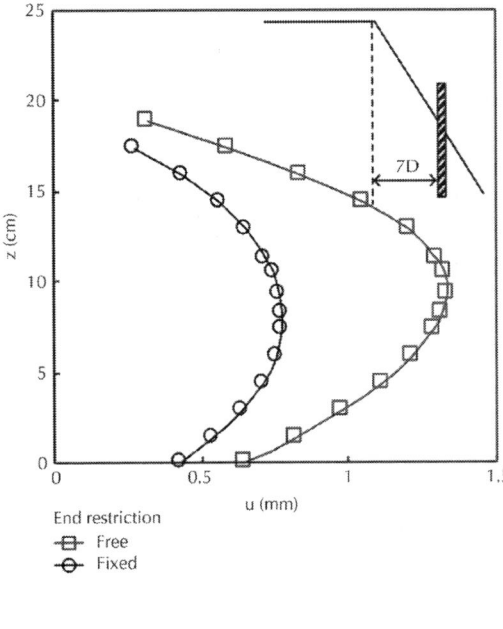

c) y=-7D

**Figure 18:** Influence of fixing styles of the pile end by numerical analysis. D: pile diameter; u: horizontal displacement.

# Inclination of Slope

The inclination of the slope was changed to 1 : 1 from the original 1.5 : 1 in the test condition, and the relative location of piles was maintained as in the original scheme. The horizontal displacements according to the different inclinations of slopes were yielded from numerical analysis (Figure 19). The horizontal displacement of the slope exhibited an evident decrease due to the decrease of inclination of slope. However, the distribution rules of horizontal displacement were consistent for the different inclinations of slope.

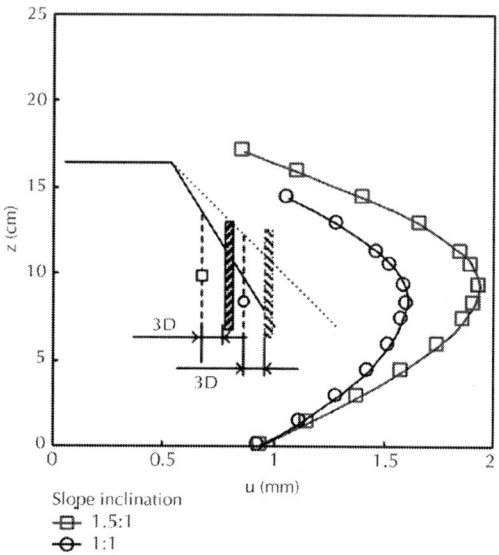

(a) 3D from pile

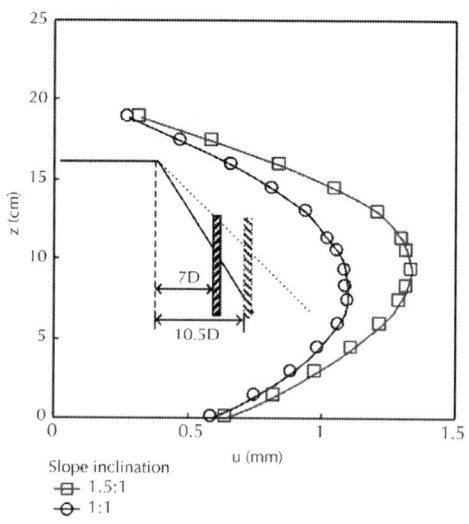

(b) y=-7D passing by slope shoulder

**Figure 19:** Influence of inclination of the slope by numerical analysis. D: pile diameter; u: horizontal displacement.

## Summary

Based on the numerical analysis, it can be concluded that the behavior of the pile-reinforced slopes was similar for the different factors considered in this paper. As discussed in above texts, the two boundary surfaces, W-surface and L-surface, can be used as important indicators of the behavior of reinforced slope. Therefore, these surfaces were summarized corresponding to different factors according to the results from numerical analysis (Figure 20). It can be seen that both of the surfaces can be described using a definite format, though their positions and curvatures were affected by different factors. Thus, it is believed that the fundamental rules and reinforcement mechanism, derived in this paper, may be suitable for most practical pile-reinforced slopes.

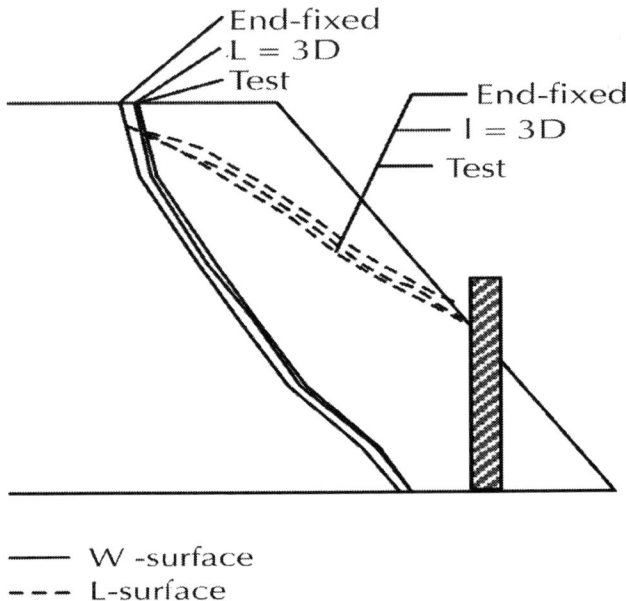

End-fixed
L = 3D
Test

End-fixed
l = 3D
Test

——— W -surface
– – – L-surface

a) Test,L=3Dpile end fixed

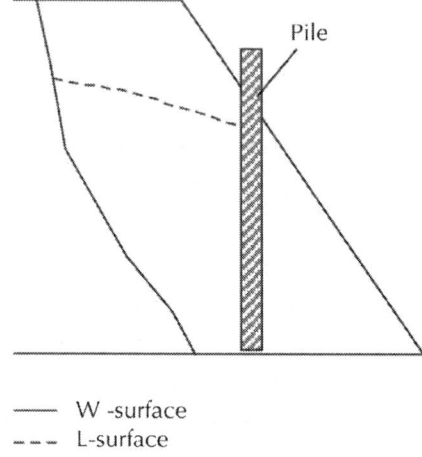

— W -surface
- - - L-surface

(b) Pile location changed

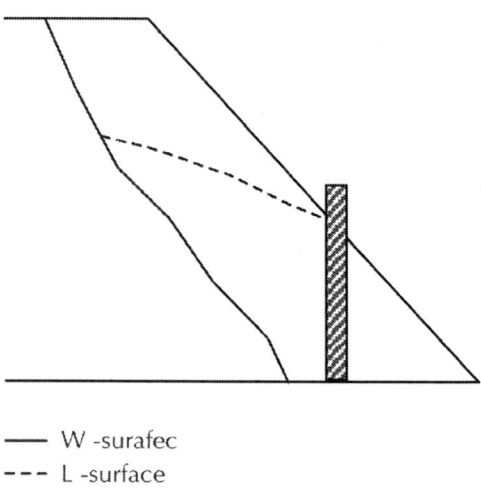

— W -surafec
- - - L -surface

(c)Slope inclination changed to 1:1

**Figure 20:** W-surfaces and L-surfaces considering different factors. L: pile spacing; D: pile diameter.

# CONCLUSIONS

The pile-reinforcement behavior of cohesive soil slope was investigated using numerical analysis and centrifuge model tests. The behavior and reinforcement mechanism of the pile-reinforced slope were obtained by considering different influence factors. The main conclusions are as follows.

1. The piles arrested a landslide that occurred in an unreinforced slope, thus, significantly increasing the stability level of the slope.

2. A numerical model was established and confirmed to be effective in capturing primary behavior of the pile-reinforced slope. Thus, a comprehensive understanding of stress-deformation response can be achieved by combining the numerical and physical simulations.

3. The piles have a significant effect on the behavior of reinforced slope; the boundary of the influenced area can be described using a continuous surface.

4. The pile-reinforcement mechanism can be described using two basic concepts: compression effect and shear effect, which, respectively, refer to the piles increasing the compression strain and decreasing the shear strain in comparison with unreinforced slope. The pile-soil interaction induces significant compression effect in the inner zone near the piles; this effect is transferred to the upper part of the slope, with the shear effect becoming prominent. As a result, possible sliding is prevented; the stability level of the slope is accordingly increased.

5. The response of pile-reinforced slope was significantly affected by several factors such as pile spacing, pile location, restriction style of pile end, and inclination of slope; however, the behavior and reinforcement mechanism are consistent for the different factors considered in this paper.

# ACKNOWLEDGMENTS

The study is supported by National Natural Science Foundation of China (nos. 51129902, 50979045) and the State Key Laboratory of Hydroscience and Engineering (no. 2013-KY-5).

# REFERENCES

1. C. D. F. Rogers and S. Glendinning, "Improvement of clay soils in situ using lime piles in the UK," Engineering Geology, vol. 47, no. 3, pp. 243–257, 1997.

2. J. A. Smethurst and W. Powrie, "Monitoring and analysis of the bending behaviour of discrete piles used to stabilise a railway embankment," Geotechnique, vol. 57, no. 8, pp. 663–677, 2007.

3. M. Kimura, S. Inazumi, J. K. A. Too, K. Isobe, Y. Mitsuda, and Y. Nishiyama, "Development and application of H-joint steel pipe sheet piles in construction of foundations for structures," Soils and Foundations, vol. 47, no. 2, pp. 237–251, 2007.

4. T. Ito and T. Matsui, "Methods to estimate lateral force acting on stabilizing piles," Soils and Foundations, vol. 15, no. 4, pp. 43–59, 1975.

5. H. G. Poulos, "Design of reinforcing piles to increase slope stability," Canadian Geotechnical Journal, vol. 32, no. 5, pp. 808–818, 1995.

6. M. A. El Sawwaf, "Strip footing behavior on pile and sheet pile-stabilized sand slope," Journal of Geotechnical and Geoenvironmental Engineering, vol. 131, no. 6, pp. 705–715, 2005.

7. M. J. Thompson and D. J. White, "Design of slope reinforcement with small-diameter piles," in Advances in Earth Structures: Research to Practice, GSP 151, pp. 67–73, ASCE, 2006.

8. A Bouafia and A. Bouguerra, "Centrifuge testing of the behaviour of a horizontally loaded flexible pile near to a slope," Canadian Geotechnical Journal, vol. 32, no. 2, pp. 324–335, 1995.

9. Porbaha and D. J. Goodings, "Centrifuge modeling of geotextile-reinforced steep clay slopes," Canadian Geotechnical Journal, vol. 33, no. 5, pp. 696–704, 1996.

10. J. G. Zornberg, N. Sitar, and J. K. Mitchell, "Performance of geosynthetic reinforced slopes at failure," Journal of Geotechnical and Geoenvironmental Engineering, vol. 124, no. 8, pp. 670–683, 1998.

11. N. I. Thusyanthan, S. P. G. Madabhushi, and S. Singh, "Tension in geomembranes on landfill slopes under static and earthquake

loading-Centrifuge study," Geotextiles and Geomembranes, vol. 25, no. 2, pp. 78–95, 2007.

12. L. Wang, G. Zhang, and J.-M. Zhang, "Nail reinforcement mechanism of cohesive soil slopes under earthquake conditions," Soils and Foundations, vol. 50, no. 4, pp. 459–469, 2010.

13. L. P. Wang and G. Zhang, "Centrifuge model test study on pile reinforcement behavior of cohesive soil slopes under earthquake conditions," Landslides, 2013.

14. T. Yamagami, J. Jiang -C, and K. Ueno, "A limit equilibrium stability analysis of slopes with stabilizing piles," in Advances in Earth Structures: Research to Practice, GSP 101, pp. 343–354, ASCE, 2000.

15. G. Zhang and L. Wang, "Stability analysis of strain-softening slope reinforced with stabilizing piles," Journal of Geotechnical and Geoenvironmental Engineering, vol. 136, no. 11, pp. 1578–1582, 2010.

16. E. Ausilio, E. Conte, and G. Dente, "Stability analysis of slopes reinforced with piles," Computers and Geotechnics, vol. 28, no. 8, pp. 591–611, 2001.

17. F. Cai and K. Ugai, "Numerical analysis of the stability of a slope reinforced with piles," Soils and Foundations, vol. 40, no. 1, pp. 73–84, 2000.

18. S. Jeong, B. Kim, J. Won, and J. Lee, "Uncoupled analysis of stabilizing piles in weathered slopes," Computers and Geotechnics, vol. 30, no. 8, pp. 671–682, 2003.

19. J. Won, K. You, S. Jeong, and S. Kim, "Coupled effects in stability analysis of pile-slope systems," Computers and Geotechnics, vol. 32, no. 4, pp. 304–315, 2005.

20. W. W. Ng and L. M. Zhang, "Three-dimensional analysis of performance of laterally loaded sleeved piles in sloping ground," Journal of Geotechnical and Geoenvironmental Engineering, vol. 127, no. 6, pp. 499–509, 2001.

21. W. W. Ng, L. M. Zhang, and K. K. S. Ho, "Influence of laterally loaded sleeved piles and pile groups on slope stability," Canadian Geotechnical Journal, vol. 38, no. 3, pp. 553–566, 2001.

22. G. R. Martin and C.-Y. Chen, "Response of piles due to lateral slope movement," Computers and Structures, vol. 83, no. 8-9, pp. 588–598, 2005.

23. G. Zhang, Y. Hu, and J.-M. Zhang, "New image analysis-based displacement-measurement system for geotechnical centrifuge modeling tests," Measurement, vol. 42, no. 1, pp. 87–96, 2009.

24. G. Zhang, D. Liang, and J.-M. Zhang, "Image analysis measurement of soil particle movement during a soil-structure interface test," Computers and Geotechnics, vol. 33, no. 4-5, pp. 248–259, 2006.

25. G. Zhang, J.-M. Zhang, and Y. Yu, "Modeling of gravelly soil with multiple lithologic components and its application," Soils and Foundations, vol. 47, no. 4, pp. 799–810, 2007.

26. G. Zhang and J.-M. Zhang, "Unified modeling of monotonic and cyclic behavior of interface between structure and gravelly soil," Soils and Foundations, vol. 48, no. 2, pp. 231–245, 2008.

27. G. Zhang and J.-M. Zhang, "Numerical modeling of soil-structure interface of a concrete-faced rockfill dam," Computers and Geotechnics, vol. 36, no. 5, pp. 762–772, 2009.

# Chapter 8

# The State-of-the-Art Centrifuge Modelling of Geotechnical Problems at HKUST

Charles W. W. Ng

Key Laboratory of Geomechanics and Embankment Engineering of Ministry of Education, Hohai University, Nanjing 210098, China.

Geotechnical Centrifuge Facility, the Hong Kong University of Science and Technology, Hong Kong, China.

## ABSTRACT

Geotechnical centrifuge modelling is an advanced physical modelling technique for simulating and studying geotechnical problems. It provides physical data for investigating mechanisms of deformation and failure and for validating analytical and numerical methods. Due to its reliability, time and cost effectiveness, centrifugemodelling has often been the preferred experimental method for addressing complex geotechnical problems. In this ZENG Guo-xi Lecture, the kinematics, fundamental principles and principal applications of geotechnical centrifugemodelling are introduced. The use of the state-of-the-art geotechnical centrifuge at the Hong Kong University of Science and Technology (HKUST), China to investigate four types of complex geotechnical problems is reported. The four geotechnical problems include correction of building tilt, effect of tunnel collapse on an existing tunnel, excavation effect on pile capacity and liquefied flow and non-liquefied slide of loose fill slopes. By reporting major findings and new insights from these four types of centrifuge tests, it is hoped to illustrate the role of state-of-the-art geotechnical centrifuge modelling in advancing the scientific knowledge of geotechnical problems.

# INTRODUCTION

In tackling some complex geotechnical problems, centrifuge modelling is often considered as a preferred experimental method. According to a survey conducted by the British Geotechnical Society in 1999, centrifuge modelling was ranked fifth in the list of the most important developments in geotechnics over the previous 50 years (Fig. 1). The ranking was based on responses from 68 geotechnical experts in academia, consulting, contracting and research organisations. It is clear from the survey that centrifuge modelling plays a key role in geotechnical engineering. In this paper, the kinematics, fundamental principles, and principal applications of geotechnical centrifuge modelling are introduced. Modelling of four complex geotechnical problems by the state-of-the-art geotechnical centrifuge at the Hong Kong University of Science and Technology (HKUST), China is described. The four geotechnical problems are: correction of building tilt, effect of tunnel collapse on an existing tunnel, excavation effect on pile capacity, and liquefied flow and non-liquefied slide of loose fill slopes. New insights from these four types of tests are revealed and the role of state-of-the-art geotechnical centrifuge modelling in improving understanding of the complex geotechnical problems is illustrated.

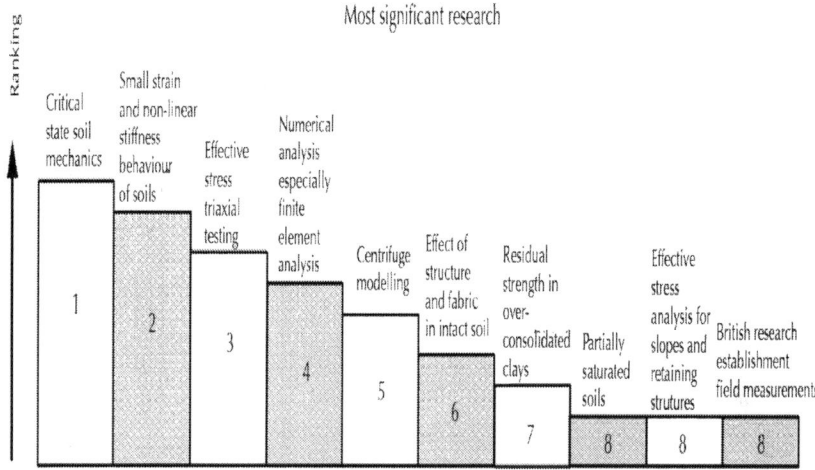

**Figure1**: The great and the good of 50 years of geotechnics (from Ground Engineering, July 1999).

# BRIEF INTRODUCTION TO THE DEVELOPMENT OF CENTRIFUGE MODELLING

According to Craig (1995), the earliest idea of using a centrifuge to increase self-weight of a small scale model was developed by Phillips in Paris in 1896. He suggested using a centrifuge to solve bridge engineering problems, but no actual test was carried out at that time. After this innovative idea, centrifuge modelling was not pursued until Bucky (1931), who conducted centrifuge model tests at Columbia University in the USA to study the integrity of mine roof structures in rock. Almost at the same time, Pokrovsky and Davidenkov from the Union of Soviet Socialist Republics (USSR) used a centrifuge to investigate problems associated with embankment and slope instability in 1933. In the subsequent two decades (1940s–1960s), a number of geotechnical centrifuges were built in the USSR and applied to tackling various problems in soils and rocks (Joseph et al., 1988). During the same period, a few research projects were undertaken in the USA by Panek (1949) and Clark (late 1950s and early 1960s). Apart from the USA and the USSR, early research work was conducted using a centrifuge by Ramberg (1968) in Sweden to study gravity tectonics, as well as by Hoek (1965) in South Africa for mining engineering. Also in the mid-1960s, the first geotechnical centrifuge was built in China by the Yangtze Water Conservancy Institute (Cheney, 1988). In 1966, the first geotechnical centrifuge in the UK was developed at Cambridge University by Prof. Schofield, who subsequently continued his work at the University of Manchester Institute of Science and Technology, where he built a large centrifuge in 1969. In the early 1970s, Profs. Rowe and Roscoe constructed centrifuges at the University of Manchester and Cambridge University, respectively. Thereafter, many researchers from various countries such as Japan, Denmark, Netherlands, and France visited Cambridge University to study centrifuge modelling techniques and to setup centrifuge facilities in their own countries (Joseph et al., 1988). In the mid-1970s, there was a renewed interest in geotechnical centrifuge modelling in the USA (Joseph et al., 1988). Centrifuge modelling was adopted to simulate geophysical events and processes by Ramberg (1968) in Chicago, craters formed by near-surface nuclear explosions and planetary impact of large bodies by Schmidt (1976), and cyclic and dynamic testing of piles by Scott (1979). After the 1980s, centrifuge modelling was recognised and well-received in many countries, especially Japan (Kimura, 1998). Since then, there has been a continuing increase in the number, size, and simulation capability of centrifuges over the world, particularly in the last ten years in China.

# KINEMATICS OF CENTRIFUGE MODELLING

Fig. 2 shows plan view of a soil model in a spinning centrifuge. In this figure, a local Cartesian coordinate system (fixed to the model container) is defined.

Consider a soil element located at an arbitrary point A. At a given time, location of the arbitrary soil element A in the model container can be expressed as a vector summation:

$$P = R + r = R_r\hat{\rho}_r + r_r\hat{\rho}_r + r_n\hat{\rho}_n \tag{1}$$

where $P$ and $R$ denote vectors from axis of the centrifuge to soil element $A$ and to the bottom of the model box (point $O$), respectively. $R$ means a vector from point $O$ to soil element $A$.

Acceleration of point $A$ is

$$\frac{d^2P}{d^2t} = \frac{d^2R}{d^2t} + \frac{d^2r}{d^2t} \tag{2}$$

By assuming the centrifuge spins at a constant angular velocity ($\omega$=constant, $\dot{\omega} = 0$ ) with a fixed radius of the centrifuge arm ($|R|$=constant), the acceleration of point $A$ can be expressed as

$$\frac{d^2P}{d^2t} = \ddot{r}_r\hat{\rho}_r + \ddot{r}_n\hat{\rho}_n - 2\omega\dot{r}_r\hat{\rho}_n + 2\omega\dot{r}_n\hat{\rho}_r - \omega^2\left(R_r + r_r\right)\hat{\rho}_r - \omega^2 r_n\hat{\rho}_n \tag{3}$$

According to their physical meanings, terms in Eq. (3) can be grouped into three parts as follows:

1.  $-\omega^2\left(R_r + r_r\right)\hat{\rho}_r - \omega^2 r_n\hat{\rho}_n$ denotes centripetal acceleration (due to spinning of centrifuge);

2.  $\ddot{r}_r\hat{\rho}_r + \ddot{r}_n\hat{\rho}_n$ describes the acceleration of a particle $P$ relative to the centrifuge platform (e.g., resulting from applied base shaking, slope failure, explosions, etc.);

3.  $-2\omega\dot{r}_r\hat{\rho}_n + 2\omega\dot{r}_n\hat{\rho}_r$ refers to Coriollis acceleration (e.g., resulting from consolidation, flow, etc.). Detailed derivations of the above equations are given by Lei and Shi (2003).

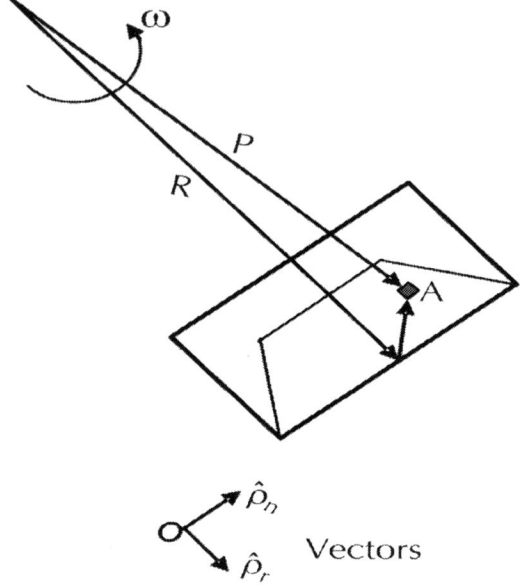

**Figure 2**: Plan view of a soil model in a spinning centrifuge.

# FUNDAMENTAL PRINCIPLES OF CENTRIFUGE MODELING

It is well recognised that soil behaviours are stress dependent. For an example illustrated in Fig. 3, a soil sample A located below the critical state line (CSL) initially, will dilate toward the CSL when it is sheared under a relatively low confining stress (e.g., in a small model test under one Earth's gravity (i.e., 1g=9.81 m/s²)). By comparison, a sample, B, having the same density (i.e., same void ratio), located at an arbitrary point above the CSL but below or on the normal compression line (NCL), will contract when it is sheared under a higher mean effective stress, $p'$ (i.e., high stress in the field or in the centrifuge). It is obvious that the use of test results from sample A for designing prototype problems is likely to be non-conservative and maybe even dangerous because the observed dilative behaviour at low stress under 1g conditions will not occur under high stress in the field. Thus, it is vital to simulate the stress level of the soil correctly before carrying out any physical experiment.

The fundamental principle of centrifuge modelling is to recreate stress conditions, which would exist in a prototype, by increasing $n$ times the "gravitational" acceleration in a $1/n$ scaled model in the centrifuge. Stress replication in the $1/n$ scaled model is approximately achieved by subjecting model components to an elevated "gravitational" acceleration, which is provided by centripetal acceleration ($r\omega^2=ng$), where $r$ and $\omega$ are the radius and angular velocity of the centrifuge, respectively. Thus, a centrifuge is suitable for modelling stress-dependent geotechnical problems. Apart from the ability to replicate in-situ stress level in a reduced size model in a centrifuge, one of the side benefits of centrifuge modelling is that the use of a small scale model shortens drainage paths of soil, resulting in a significant reduction of consolidation time by $1/n^2$.

For centrifuge model tests, scaling laws are generally derived through dimensional analysis, from the governing equations for a phenomenon, or from the principles of mechanical similarity between a model and a prototype (Taylor, 1995; Garnier et al., 2007). Some common scaling factors derived and used are summarised in Table 1.

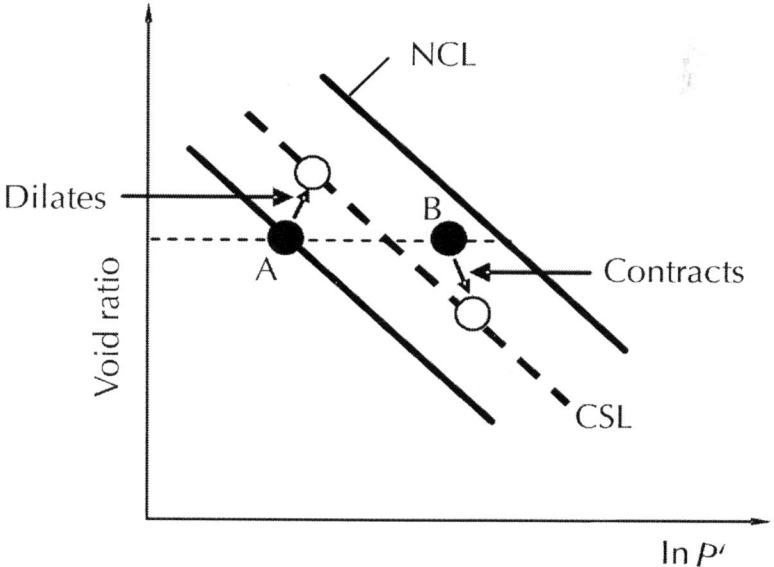

**Figure 3**: Distinct responses of two soil samples at the same density sheared under different confining stresses.

**Table 1**: Some common scaling factors for centrifuge tests

| Parameter | Scale factor (model/prototype) |
|---|---|
| Acceleration | $n$ |
|  |  |
| Linear dimension | $1/n$ |
| Stress | 1 |
| Strain | 1 |
| Mass | $1/n^3$ |
| Density | 1 |
| Unit weight | $n$ |
| Force | $1/n^2$ |
| Bending moment | $1/n^3$ |
| Bending moment/unit width | $1/n^2$ |
| Flexural stiffness | $1/n^4$ |
| Flexural stiffness/unit width | $1/n^3$ |
| Time (dynamic) | $1/n$ |
| Time (consolidation/diffusion) | $1/n^2$ |
| Time (creep) | 1 |
| Pore fluid velocity | $n$ |
| Velocity (dynamic) | 1 |
| Frequency | $n$ |

# PRINCIPAL APPLICATIONS OF CENTRIFUGE MODELLING

According to Ko (1988), four principal applications of geotechnical centrifuges can be classified as follows.

## Modelling of Prototype

Modelling of prototype is an obvious and direct application of the centrifuge modelling technique to simulate and tackle actual engineering problems. Some common applications include investigating slope instability, pile capacity, and the effect of tunnelling/excavation on adjacent existing underground structures. Both qualitative and quantitative analyses are possible from model tests.

# Investigation of New Phenomena

Centrifuge modelling has been successfully applied to the study of various unusual phenomena that are not well understood and are extremely difficult to study. Typical examples include plate tectonics, crater formations by nuclear explosions, various earthquake-induced events and soil liquefaction, and transportation of contaminants in soil. Behaviour of loose-fill slopes subjected to various rainfall and earthquake conditions can also be investigated.

# Parametric Studies

Parametric study in geotechnical centrifuge modelling is an example where physical model experiments are best rewarded. Normally, a major effort is necessary to design and manufacture the first model, while the actual testing and small variations in the model are relatively easily performed. By varying some model parameters (geometry, loading and boundary conditions, rainfall intensity or soil type), the sensitivity of test results to these variations can be evaluated and the most critical parameters can be identified. This leads directly to the possibility of generating useful design charts. Examples include bearing capacity of footings on slopes, critical design parameters in flow processes, and capacity of laterally loaded pile groups.

# Validations of Numerical Methods

Any modelling technique, either physical or numerical, demands the acceptance of simplifications and assumptions. In many cases, numerical techniques are still limited to 2D problems for various reasons, but centrifuge modelling does not impose this restriction. It is often easier to simulate a 3D than a 2D plane strain problem in a centrifuge. For investigating any complicated geotechnical problem, it would be ideal to perform both numerical analyses and centrifuge model tests. The results from these two techniques can then be compared and verified (discussed later).

# THE STATE-OF-THE-ART GEOTECHNICAL CENTRIFUGE AT HKUST

One of the most advanced geotechnical centrifuges in the world was established at HKUST in April 2001 (Ng et al., 2001a), as shown in Fig. 4. This 400g-t geotechnical centrifuge is equipped with advanced simulation capabilities including the world's first in-flight bi-axial (2D) shaker, an advanced four-axis robotic manipulator and a state-of-the-art data acquisition and control system. Figs. 5 and 6 show the bi-axial shaking table (Shen et al., 1998; Ng et al., 2001a) and the four-axis robotic manipulator (Ng et al., 2002), respectively.

This 8.4 m diameter beam centrifuge is equipped with two swinging platforms, one for static tests and one for dynamic tests. For static tests, the centrifuge is able to accommodate a model size of up to 1.5 m×1.5 m×1 m. The centrifugal acceleration can be up to 150g. For dynamic tests, the centrifuge incorporates a unique bi-axial servo-hydraulic shaker (Fig. 5) to model earthquake-induced engineering problems (Ng et al., 2001a; 2004b). The bi-axial shaker is capable of simulating earthquake motions in two horizontal directions simultaneously. The shaker can accommodate a mode size of up to 0.6 m×0.6 m×0.4 m and up to 3000 N in weight. The centrifuge can be operated at up to 75g for dynamic tests.

As shown in Fig. 6, the advanced and state-of-the-art four-axis robotic manipulator has incorporated a tool changer and four tool adopters to permit interchanging tools without stopping the centrifuge. At a centrifugal acceleration of 100g, the robotic manipulator can produce a torque up to ±5 MN·m and prototype loads of ±10 MN, ±10 MN, and 50 MN forces in the $x$, $y$, and $z$directions, respectively.

Fig. 7 compares the capacity of major geotechnical centrifuges worldwide. Acronyms of some institutions in the figure are given in

Appendix.

It can be seen that the capacity of the centrifuge at HKUST (400g-t) is one of the largest in the world. This centrifuge has been used to model and investigate various complex and challenging geotechnical problems. Four examples are reported and interpreted in the following sections.

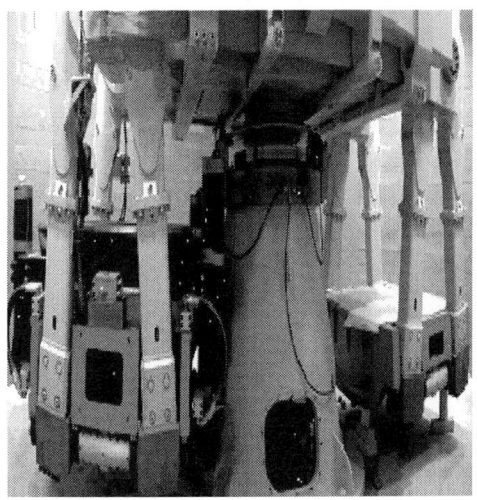

**Figure 4**: The 8.5 m-diameter (400g-t) beam centrifuge at HKUST.

**Figure 5**: The bi-axial shaking table at HKUST.

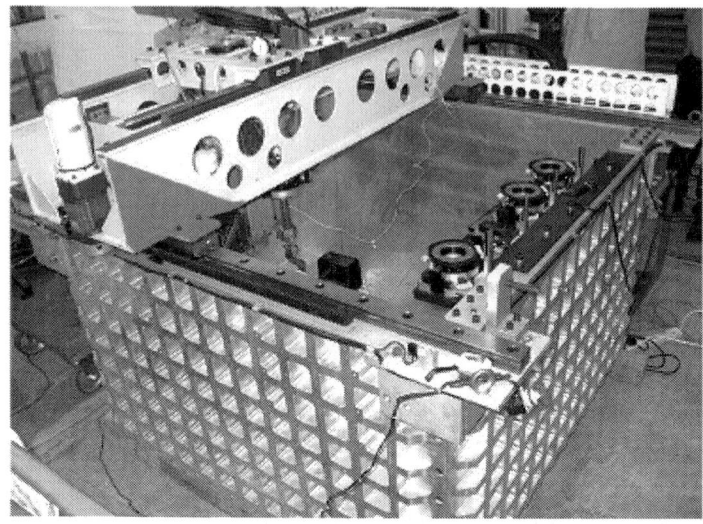

**Figure 6**: The four-axis robotic manipulator at HKUST.

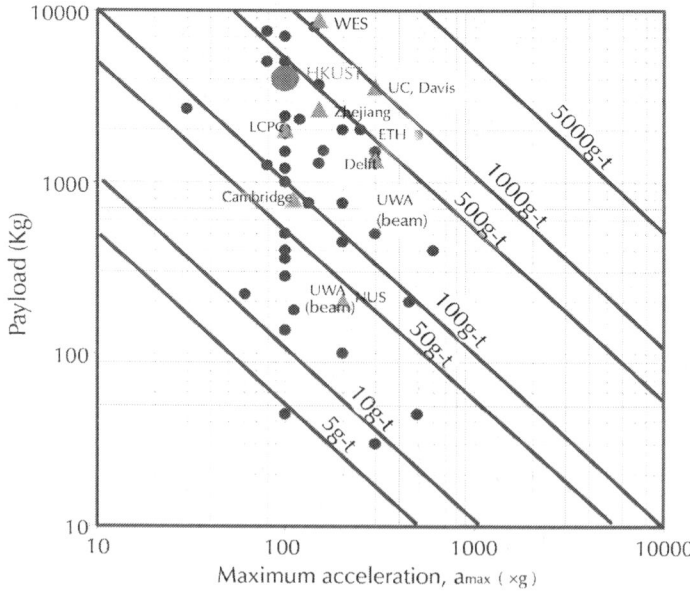

**Figure 7**: Capacity of the major geotechnical centrifuges in the world.

# Example 1: Correction of Building Tilt by in-Flight Soil Extraction

## Introduction

Building tilt is frequently encountered where ground is not homogeneous or there are adjacent underground constructions conducted. To correct building tilts, some solutions have been proposed. The Italian engineer Terracina (1962) proposed one possible way to correct the Leaning Tower of Pisa, by extracting soil through inclined boreholes from the less settled side of the leaning tower. Similar to soil extraction through inclined boreholes, the use of relief boring techniques with soil extraction through vertical boreholes is popular in China to reduce tilting and to stabilise buildings on soft ground. It is recognised that vertical soil extraction is generally simpler than an inclined extraction underneath the foundation of a tilted building. However, the effectiveness of vertical boring, as compared with inclined drilling, is somewhat doubtful and is questioned by some engineers. Sometimes the design of remedial action to reduce building tilt is essentially empirical. To investigate the effectiveness of vertical soil extraction, in-flight vertical boring adjacent to an initially tilted building was simulated, by the four-axis robotic manipulator at HKUST. In addition, results of the centrifuge test were compared with predictions from a theoretical elastic solution based on Mindlin (1936)'s equations (Ng and Xu, 2003).

## Test Setup, Model Preparation, and Procedure

Fig. 8 shows the test setup as reported by Ng and Xu (2003).

The test was conducted at 30g. The simulated building had a 5.4 m×5.4 m base area (in prototype) and was approximately nine-storeys tall (in prototype), which generated an average bearing pressure of 89 kPa on the ground. The model ground consisted of an unsaturated completely decomposed granite (CDG) prepared at an average initial water content of 15.3% and a dry unit weight of 13.0 kN/m$^3$. A hollow cylinder extracting tool with an external diameter of 30 mm and a soil bin were developed for the test. The movements of the building were measured by two linear variable differential transformers (LVDTs) and one non-contact laser displacement transducer. The performance of the robot and the model was monitored by five cameras at different locations.

When the centrifuge was spun up to 30g, the building was at an initial tilt of 1/27. To correct the tilted building, some soil was extracted from the ground in the opposite side of the tilt by drilling two series of holes. The sequence of drilling hole is shown in Fig. 8b. The holes were 160 mm deep and 100 mm away from the building. Each hole was drilled in two steps, each 80 mm in depth. Converted to prototype scale, each hole was 0.9 m in diameter and 4.8 m deep, and was 3 m away from the building.

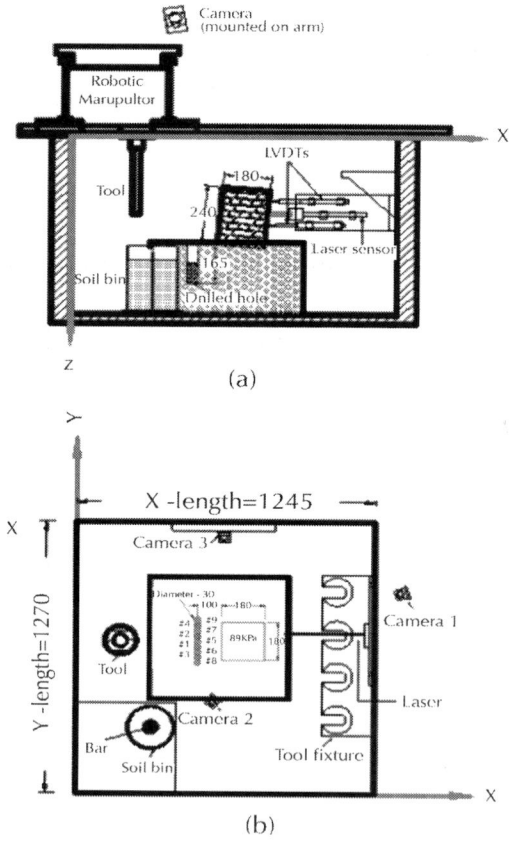

**Figure 8**: Centrifuge test setup: (a) elevation view; (b) plan view (all dimensions in mm, unless stated otherwise).

## Effects of Soil Extraction on Building Tilt

Fig. 9 shows the development of building tilt during soil extraction. The building tilt calculated by a theoretical elastic solution is also included for comparison (Ng and Xu, 2003). The measured results show that drilling of the first series of holes (holes #1–4) caused a 1.5% reduction of the building tilt while the second series of holes (holes #5–9) caused a further 0.7% correction to the building.

After drilling nine holes, the final tilt of the building was reduced to 1.5%. This test clearly demonstrated the effectiveness of correcting building tilt by vertical soil extraction.

Based on Mindlin (1936)'s equation, Ng and Xu (2003) derived an elastic solution and compared it with the centrifuge test results by soil extraction. As expected, the calculated change of building tilt (Fig. 9) shows a similar trend to the measured building tilt, but with a much smaller magnitude (only 45% of the measured value at the final stage). Two major reasons for the underestimation are:

1.    The solution is derived for vertical boring of a single hole, so the influence of a previous boring on a subsequent one was not taken into account.

2.    Yielding and plastic deformation of the soil were not considered.

Therefore, the construction sequence and the plastic deformation of soil should be taken into account when predicting the correction of building tilt.

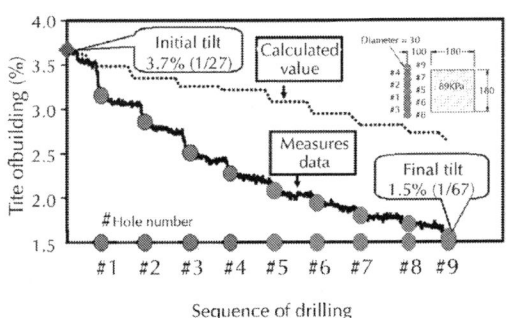

**Figure 9:** Development of tilting of model building during soil extraction (all dimensions in mm, unless stated otherwise).

# Example 2: Effects of Collapse of a Tunnel on an Existing Tunnel

## Introduction

Rapid development in urban areas leads to a high demand for transportation of people and vehicles. Tunnels are often constructed to reduce congestion and to serve as vital conduits for the movement of vehicles. Sometimes, it is inevitable that multiple tunnels are constructed very close together due to intensive use of underground space (Ng et al., 2004a). Tunnel excavation will result in volume loss which induces movement and stress change in the soil around adjacent tunnels. An extreme case of tunnel-tunnel interaction problem would be the influence of a tunnel collapse on an adjacent tunnel. One typical example is the collapse of the Heathrow Express Tunnel in 1994 when the collapse of three tunnels occurred during the construction of a new tunnel in London Clay (HSE, 2000). Therefore, an understanding of the effects of multiple tunnel interaction is essential to develop an improved design guide so that precautions can be taken during and after new tunnel construction. The effect of tunnel collapse on a nearby existing tunnel has been investigated by carrying out centrifuge tests (Ng et al., 2003).

## Centrifuge Model Package

Figs. 10a and 10b show elevation view and bird's eye view of the model package, respectively (Ng et al., 2003). The test was carried out at 60g. The existing model tunnel was made of an aluminium-alloy pipe with an external diameter of 150 mm, wall thickness of 3.3 mm, and longitudinal length of 1243 mm. The estimated $EI$ in transverse section per metre run was 0.21 kN·m$^2$, which was used to represent a prototype reinforced concrete tunnel of an external diameter of 9 m, wall thickness of 0.268 m, and a length of 74.6 m. The new tunnel with prototype diameter of 9 m and length of 27 m was designed to simulate an open-face excavation similar to an idealised New Austrian Tunnelling Method (NATM). The idealised NATM was simulated by using 12 semi-circular airbags, i.e., 12 compartments (Fig. 11). Each semi-circular airbag was 150 mm in diameter and separated by 3 mm thick aluminium-alloy dividers. Each pair of semi-circular airbags was used to mimic the left drift and right drift of a NATM tunnel. The advance of the new tunnel was simulated by reducing the air pressure inside the airbags at the construction stages indicated in Fig. 11.

Dry Leighton Buzzard (LB) sand fraction E was used in the test. The maximum and minimum unit weight of the sand was equal to 15.9 kN/m³ and 13.2 kN/m³, respectively. In this centrifuge test, the unit weight of the sand was 14.2 kN/m³ (i.e., relative density=50%).

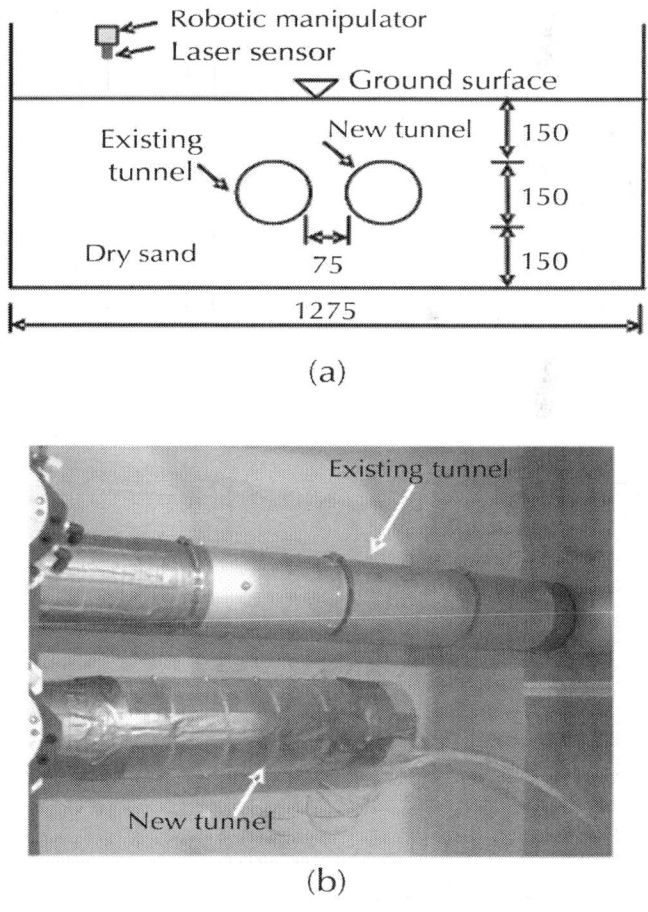

**Figure 10:** Centrifuge model package: (a) elevation view (unit: mm); (b) bird's eye view (unit: mm).

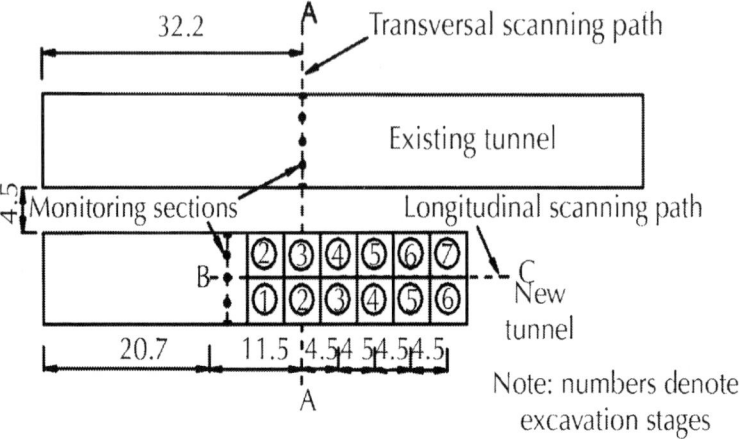

**Figure 11**: Layout of tunnels and monitoring sections (unit: m).

## Instrumentation and Testing Procedure

Fig. 11 shows the plan view of the twin tunnels and two monitoring sections in prototype scale (Ng et al., 2003). In each monitoring section, eight miniature pressure cells and eight pairs of strain gauges were installed evenly with an interval of 45° around the circumference of each tunnel. The pressure cells were mounted on the outer surfaces and the strain gauges were glued onto both the outer and inner surfaces of the tunnels to measure circumferential distributions of axial strain and bending strain. Three-dimensional ground surface profiles were measured by using a laser sensor mounted on the robotic manipulator in-flight (Fig. 10a). Two travelling and scanning paths along sections AA and BC by the robotic manipulator were adopted and controlled by a computer.

During the swing-up of the centrifuge, the air pressure inside the airbags was increased to balance the increased overburden pressure as the centripetal acceleration was increased. When the centripetal acceleration reached 60g, the construction of the new tunnel was started by depressurising the airbags in different stages to stimulate an idealised NATM. The construction sequence was divided into seven stages starting from the right drift of the tunnel, as indicated by the stage numbers shown in Fig. 11.

## Three-Dimensional Ground Surface Settlements and Bending Moments

Figs. 12a and 12b show the measured settlements along the transverse path AA and the longitudinal path BC from excavation stage 1 to stage 3, respectively. The results are presented in prototype scale unless stated otherwise.

After excavation stage 1, a settlement trough formed at section AA with the maximum settlement above the right drift of the tunnel, as expected. As the tunnel advanced, the transverse settlement trough at the section became deeper and wider with the location of the maximum settlement shifted towards the centreline of the tunnel. By fitting a Gaussian distribution with the maximum settlement of 260 mm and the point of inflection of the settlement trough equal to 5.5 m, it can be seen that there was a considerable discrepancy between the fitted distribution and the measured settlements above the left drift of the tunnel, possibly attributable to the presence of the existing tunnel and the excavation sequence adopted. In the longitudinal direction, the measured settlements increased with the advance of the tunnel as expected and the measured maximum settlements were consistent with those along the transverse section and a 3D settlement bowl can be deduced.

Fig. 13 shows the measured bending moments (BM) normalised by the ultimate moment capacity, $M_u$. For the model tunnel with an estimated yield stress of 100 MPa, the calculated ultimate moment capacity was $M_u$=0.272 kN·m/m. The maximum positive BM was induced at the crown ( =0°) and at the invert ( =180°) of the existing tunnel. As expected, the BM increased as the tunnel advanced, particularly at the crown. On the other hand, the maximum normalised negative BM occurred at the left and the right springlines (i.e., =90° and =270°, respectively). The BM at =270° increased steadily whereas the BM at =90° remained almost unchanged. Based on the measurements, it can be deduced that the tunnel deformed into an elliptical shape. The absolute increase in BM due to tunnelling was 60%, 28, and 228% at the crown, invert, and the right springline of the existing tunnel, respectively.

Recently, Ng et al., (2013a) have also reported a series of 3D centrifuge tests investigating the responses of an existing tunnel to the excavation of a new tunnel perpendicularly underneath it. 3D tunnel advancement was simulated using a novel technique that considers the effects of both volume and weight losses in-flight. This novel technique involves using a "donut" to control volume loss and mimic soil removal in-flight. To improve fundamental understanding of stress transfer mechanism, measured results were back analyzed three-dimensionally using the finite element method using an advanced soil model, which can capture soil behaviour at small strains. It is found that the maximum measured settlement of the existing tunnel induced by the new tunnel constructed underneath was about 0.3% of tunnel diameter, which may be large enough to cause serviceability problems. The observed large settlement of the existing tunnel was caused not only by a sharp reduction in vertical stress at the invert but also by substantial overburden stress transfer at the crown. The section of the existing tunnel directly above the new tunnel was vertically compressed because the incremental normal stress on the existing tunnel was larger in the vertical direction than in the horizontal direction. The tensile strain and shear stress induced in the existing tunnel exceeded the cracking tensile strain and allowable shear stress limit given by the American Concrete Institute.

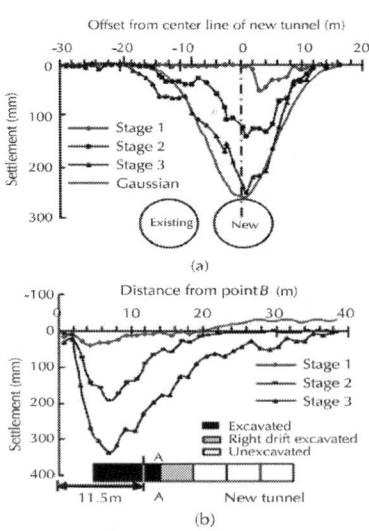

**Figure 12**: Three-dimensional settlement bowl along transverse direction (a) and longitudinal direction (b) of the new tunnel.

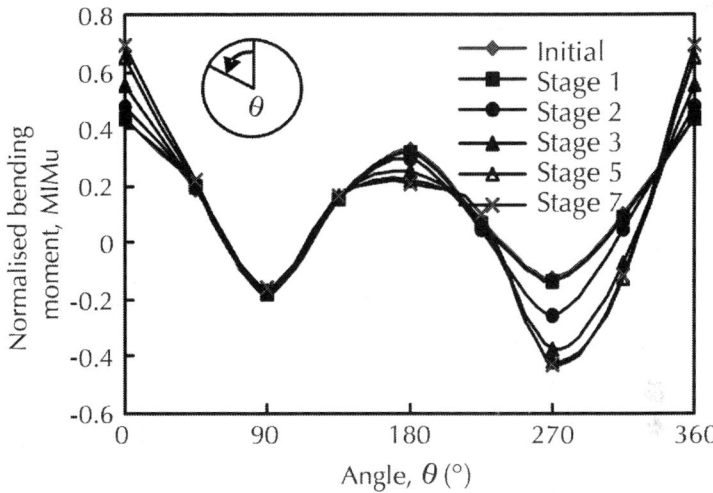

**Figure 13**: Normalised bending moment distribution.

# Example 3: Excavation Effects on Pile Capacity

## *Introduction*

Conventionally, pile loading tests are carried out at the ground surface, prior to a basement excavation. A sleeve is often used in the loading test to eliminate shaft resistance within the depth of future excavation, in order to predict the behaviour of the piles working underneath the basement. Following such a test procedure, the stress changes in the soil due to excavation are not captured. Predicting the performance of piles beneath a basement from the conventional loading tests may pose a challenge for engineers.

In order to study the influence of excavation on pile capacity, centrifuge modelling of single pile loading tests was carried out in dry Toyoura sand (Zheng et al., 2010; 2012). In-flight pile loading tests were carried out both at the ground surface prior to excavation and at the formation level after excavation. Model piles with two distinct interfaces (i.e., low friction and high friction) were also simulated to investigate the influence of interface roughness on pile behaviour and capacity.

## Model Setup and Testing Procedures

Fig. 14a shows a schematic diagram of the tests which are intended to model conventional pile loading tests. In-flight loading tests were carried out at the ground surface. A sleeve was used for the upper portion of the pile to eliminate shaft resistance within the depth of future excavation. Fig. 14b shows the tests which are intended to model pile loading tests carried out after excavation.

In-flight basement excavation was simulated by draining away zinc chloride solution, which had the same density as that of the sand. A circular model diaphragm wall was used to retain the excavation. The lateral stiffness of the diaphragm wall was very large, and hence the lateral deformation caused by the excavation was negligible. After excavation, an in-flight pile loading test was carried out at the basement level. In each of the tests, an instrumented single pile was located at the centre of the excavation area. All the centrifuge tests were performed at 100g.

Model piles with two distinct interfaces were used. They were defined as "low friction piles" and "high friction piles" (Zheng et al., 2012). Normalised roughness $R_n$ (Kishida and Uesugi, 1987) of the two interfaces were 0.018 and 0.21, respectively. According to laboratory tests (Kishida and Uesugi, 1987; DeJong and Frost, 2002; Fioravante, 2002), for the low friction piles under loading, particle sliding failure happens at the soil-pile interface. These low friction piles are therefore intended to simulate piles in non-dilatant geo-materials (such as normally consolidated clay and loose sand). In contrast, for the high friction piles, failure happens within the soil surrounding the piles. Volume change takes place in the shear band according to the density and the stress level of the soil. The high friction piles are therefore intended to simulate piles in dilatant geo-materials (such as dense sand).

The pluvial deposition method was used to rain sand into the model container from a hopper. After the formation of sand bed, the model pile was temporarily fixed in its designed location. Then the process of sand raining was continued. By using the pluvial deposition method to form sand bed around the pile at 1g, the initial stress around the model pile is small. Subsequently, as the g-level increased during a test, the initial stress around the pile also increased under the $K_o$

conditions, which could be regarded as similar to that adjacent to a non-displacement pile. The measured average relative densities of the sand in the four tests ranged from 58% to 64%, with an average value of 62%.

All the test results are converted to prototype scale unless stated otherwise.

(a)

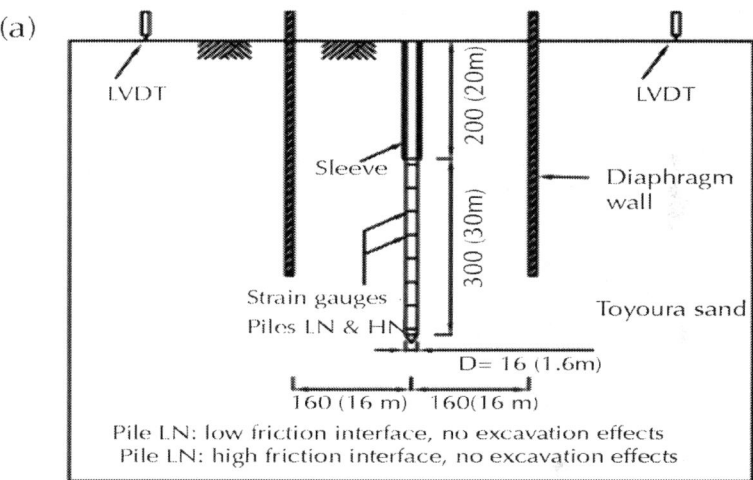

(b)

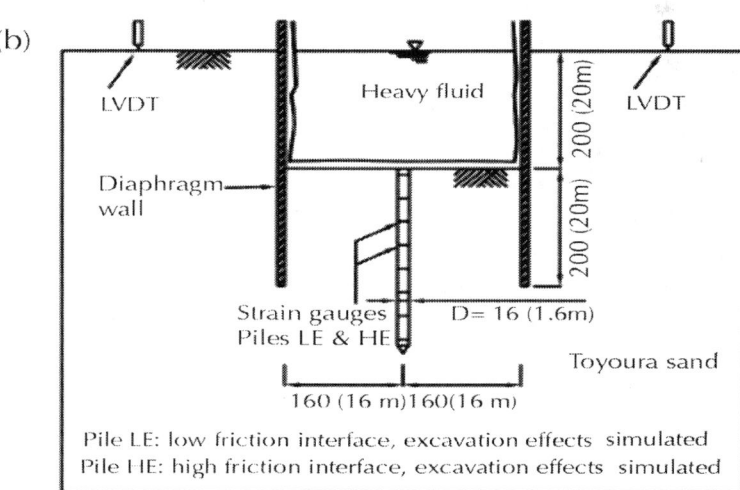

**Figure 14**: Schematic diagrams of the centrifuge models: (a) piles tested at ground surface (LN and HN); (b) piles tested after excavation (LE and HE) (Units: mm. Numbers in parenthesis denote corresponding prototypes).

# Influence of Excavation on Pile Performance and Capacity

Fig. 15a shows the normalised load-settlement relationships of the low friction piles. To compare the ultimate pile capacities, the failure criterion proposed by Ng et al. (2001b) is used. This is a semi-empirical method for the interpretation of a moderately conservative failure load. The criterion for floating piles is given by

$$\Delta M = 0.045D + \frac{1}{2}\frac{PL}{AE}$$ (4)

where $_M$ is the maximum pile head movement to define the ultimate load; $D$ is the pile diameter; $P$ is the applied test load; $L$ is the pile length; $A$ is the pile shaft cross-sectional area; and $E$ is the pile shaft modulus of elasticity.

As shown in the figure, the interpreted capacities of the two 1.6 m-diameter piles (LN and LE) are 14 700 kN and 11 600 kN, respectively. The capacity of pile LE is 80% of that of pile LN. For comparison, the failure criterion based on a pile settlement of 10%$D$ is also illustrated in the figure. If this criterion is adopted, the capacity of pile LE is 84% of that of pile LN. It is evident that the capacity of a low friction pile is reduced by 16%–20% after excavation due to the vertical stress relief.

Fig. 15b shows the load-settlement relationships of the high friction piles (Zheng et al., 2012). As an incremental load is applied, the pile tested after excavation (HE) has slightly higher stiffness than that of pile HN. Using the same failure criterion (Ng et al., 2001b), interpreted capacities for piles HN and HE are 16 960 kN and 27 680 kN, respectively. The capacity of pile HE is 39% higher than that of pile HN. Based on these results, the capacity of a high friction pile is increased after excavation. This finding is opposite to the results for the low friction piles. The increase in pile capacity after excavation is probably attributable to the increase in horizontal stress resulting from strong soil dilation at the high friction pile-soil interface.

Fig. 16a shows the mobilisation of the average unit shaft resistance for the low friction piles (Zheng et al., 2010). The shaft resistance of each pile is calculated by subtracting the toe resistance from the total load applied at the pile head. At the final loading stage, the average unit shaft resistance of pile LN is 95 kPa. In contrast, the unit shaft resistance of pile LE is only 38 kPa, which is 40% of that of pile LN. The reduction in the unit shaft resistance is closely related to the change in the effective stress level for the two cases. It is found that the ultimate unit shaft resistance of a low friction pile decreases in proportion to the effective stress relief resulting from excavation.

Fig. 16b shows the average unit shaft resistance for the high friction piles (Zheng et al., 2012). At the final loading stage, the average unit shaft resistance of pile HE is about 272 kPa, which is 26% higher than that of pile HN (216 kPa). This result is consistent with the higher capacity of pile HE, as shown in Fig. 16b. This measured increase in unit shaft resistance may be due to the increase in horizontal stress resulting from the dilative behaviour of the soil-pile interface.

Recently Ng et al. (2013b) have also reported two 3D centrifuge tests investigating the effects of a basement excavation on an existing tunnel. In addition, a preliminary 3D numerical analysis was conducted to back-analyse the centrifuge tests and to study the effects of the tunnel cover-to-diameter and unloading ratios on the existing tunnel. For the specific conditions simulated and soil type tested, a maximum heave of about 0.07% of the final depth of the basement excavation ($H_e$) was induced in the tunnel that ran parallel to and beneath the basement. On the contrary, a maximum settlement of 0.014% $H_e$ was induced in the tunnel located at the side of the basement. For the former tunnel, the influence zone by the basement excavation on vertical tunnel displacement along the longitudinal direction was 1.2$L$ (basement length). By studying the measured strains in the longitudinal direction of the existing tunnel, it was found that the inflection point, where the shear force is at a maximum, was located at 0.8$L$ away from the basement centre. Due to stress relief from the basement excavation, the tunnel that located directly beneath the basement was vertically elongated but the one that lay at the side of the basement was distorted. A preliminary numerical parametric study found that tunnel heave decreased as the cover-to-diameter ratio increased but at a reduced rate.

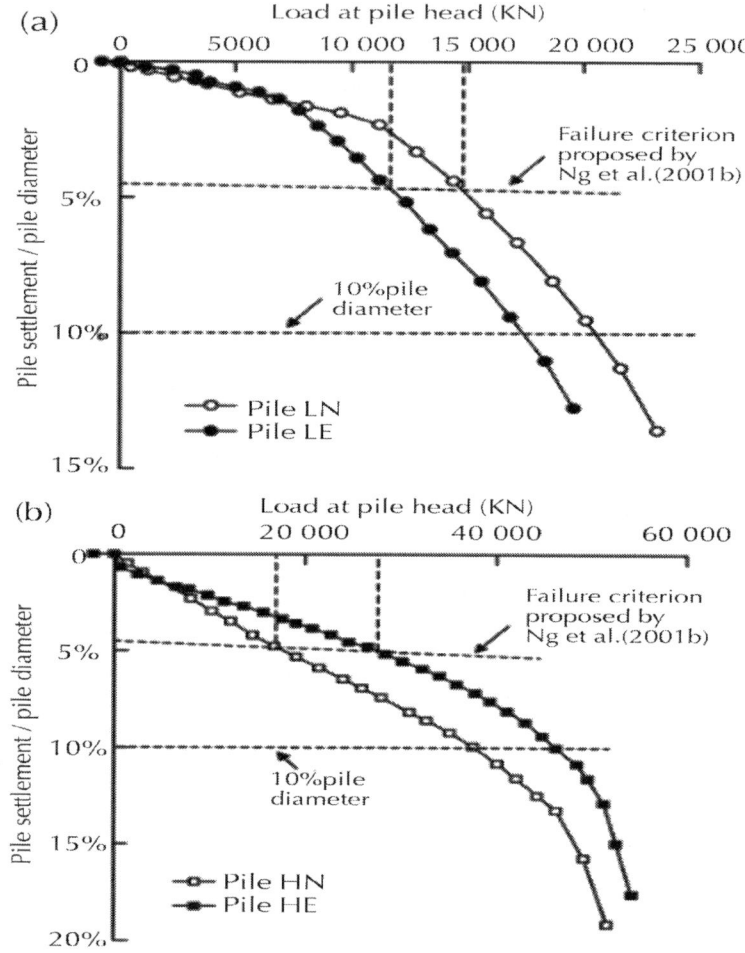

**Figure 15**: Load-settlement curves: low friction piles (LN and LE) (a); high friction piles (HN and HE) (b).

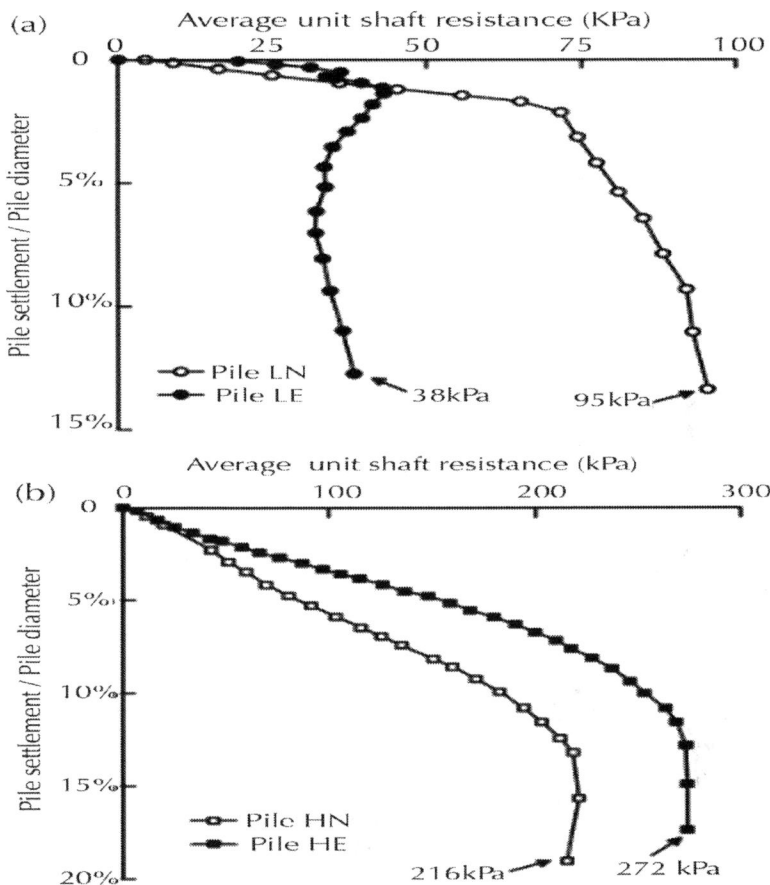

**Figure 16**: Average unit shaft resistances: low friction piles (LN and LE) (a); high friction piles (HN and HE) (Zheng et al., 2012) (b).

# Example 4: Liquefied Flow and Non-Liquefied Slide of Loose Fill Slopes

## Introduction

Slope failures occur in many parts of the world. A slope will become unstable when its shear resistance is smaller than any external driving shear stress, which may be induced by mechanical and hydraulic means such as rainfall, earthquake, vibration, and seepage. Alternatively, a slope will also become unstable if its shear resistance has deteriorated and reduced due to weathering and any other mechanisms such as static liquefaction. Very often the terminology "static liquefaction" is used to describe soil slope failures and is reported in the literature. However, it is evident that different researchers and engineers are referring to different failure mechanisms. Some use debris mobility (travel angle or run out distance) to judge whether a slope failure has been caused by liquefaction or not. Clearly there is no direct relationship between liquefaction and mobility. For instance, level ground can liquefy (at zero/small effective stress under seismic loading) with zero run out distance. On the contrary, a steel ball can run down a bare slope and travel a long way but that has nothing to do with liquefaction (Ng, 2009).

What is static liquefaction? How is it triggered? What is the effective stress at failure, if the slope is fully saturated initially, as in an undersea slope? How can we identify and define static liquefaction failures? Does a strain-softening material necessarily mean static liquefaction? Is there any difference between slide failure and flow failure? What is the role of hydrofracture? How does the angle of a slope affect the so-called static liquefaction? Is there any difference between fluidisation and liquefaction? Will static liquefaction occur in unsaturated soil slopes? How does the angle of a slope affect the potential of static liquefaction? Is there any relationship between the so-called static liquefaction failure and run out distance? Can soil nails be used to stabilise any loose fill slopes? Some of these questions have not been well understood and addressed and some of them may be even controversial. Some selected issues discussed above were investigated by Ng (2005; 2007; 2008) by means of laboratory triaxial element tests and centrifuge model tests on loose fill slopes using gap-graded LB sand and a well-graded silty sand (i.e., CDG). Observed key failure mechanisms of static liquefaction in the LB sand and non-liquefied slides of CDG fill slopes are identified and discussed.

## Clarification of Some Terminologies Relating to Static Liquefaction

Fig. 17 shows some typical results from undrained monotonic loading triaxial tests on saturated, anisotropically consolidated sand specimens.

As illustrated, a very loose sand specimen, A, exhibits a peak undrained shear strength at a relatively small shear strain and then "collapses" to much smaller shear strength at large strains. This behaviour is often casually referred to as "liquefaction" or "flow liquefaction" by many researchers and engineers. No matter whether it is called "flow liquefaction" or "liquefaction", the terminology used to describe the behaviour observed in the laboratory is rather confusing and, strictly speaking, incorrect. Would it be clearer and more precise to describe the material behaviour of the loose specimen, A, and a dense specimen, B, as "strain-softening" and "strain-hardening", respectively? It must be pointed out that these are just the behaviour of the material element and do not necessarily capture and represent the global behaviour of an entire fill slope or an earth structure.

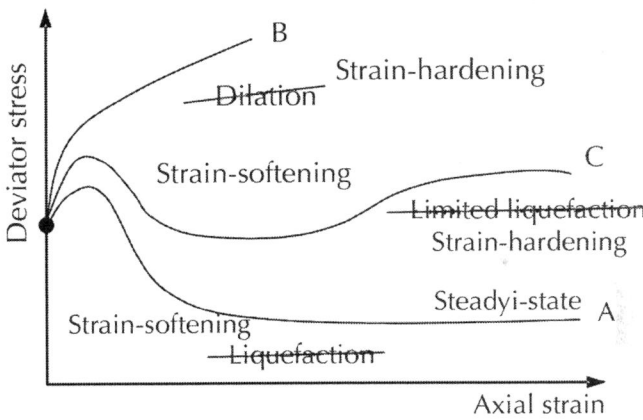

**Figure 17**: Incorrect interpretations of liquefaction, limit liquefaction and dilation in undrained monotonic loading tests on sand (modified from Castro (1969) and Kramer (1996)).

## *Failure Mechanism of Liquefied Flow in Sand Fill Slopes*

Slope centrifuge model tests were carried out to investigate the failure mechanisms of static liquefaction of loose fill slopes subjected to a rising ground water table (Zhang et al., 2006; Ng, 2009). LB Fraction E fine sand was selected as the fill material for the model tests, because of its pronounced strain-softening characteristics and its high liquefaction potential, i.e., a substantial reduction in shear strength when it is subjected to undrained shearing (Cai, 2001; Zhang, 2006). Fig. 18 shows the gap-graded particle size distribution of LB sand. D10 and D50 of the sand were 125 m and 150 m, respectively. Following BS1377 (1990), the maximum and minimum void ratios of the LB sand were found to be 1.008 and 0.667, respectively (Cai,2001). The estimated saturated coefficient of permeability was $1.6 \times 10^{-4}$ m/s.

Fig. 19 shows an instrumented 29.4° loose sand fill slope model together with the locations of the pore water pressure transducers (PPTs) (Zhang and Ng, 2003; Ng, 2008). The model slope was prepared by moist tamping. The initial relative compaction was 68%.

The body of the sand slope was instrumented with seven PPTs and arrays of surface markers were installed for image analysis of soil movements. LVDTs and a laser sensor were mounted at the crest of the slope to monitor its settlement.

Although the initial angle of the loose slope was prepared at 29.4° at 1g, the slope was densified to 80% of the maximum relative compaction due to self-weight compaction at 60g. The slope angle was therefore flattened to 24° (Fig. 20a), which is steeper than the angle of instability of 18.6°. This implies that the slope was vulnerable to instability, which could lead to liquefaction. At 60g, the 18 m-height (prototype) slope was destabilised by rising ground water from the bottom of the model (Zhang, 2006). The flow rate of the raising ground water was 9.6 mm/h in the model scale, corresponding to 573 mm/h in the prototype (based on a scaling factor of $n$ for pore fluid velocity, Table 1).

The loose sand slope liquefied statically and flowed rapidly (Fig. 20b), i.e., it followed a process in which the loose slope was sheared under undrained conditions, lost its undrained shear strength as a result of the induced high pore water pressure and then flowed like a liquid, called "liquefied flow".

Fig. 21 shows the measured rapid increase in the excess pore water pressure ratio ( $u/{'}_v$ ) within about 25 s (prototype) at failure at a number of locations in the slope during the test. The maximum measured $u/{'}_v$ was about 0.6, which would be much higher if a properly scaled viscous pore fluid were used to reduce the rate of dissipation of excess pore pressure in the centrifuge. This means that the slope would liquefy much more easily. As shown in Fig. 20b, the completely liquefied slope inclines at about 4° to 7° to the horizontal after the test. The observed fluidisation from in-flight video cameras and the significant rise in excess pore water pressures during the test clearly demonstrated the static liquefaction of the loose sand fill slope. It should be noted that measurements of sudden and significant rise of excess pore water pressures are essential to "prove" or verify the occurrence of static liquefaction of loose fill slopes if no video recording is available. The liquefaction of the loose sand slope was believed to be initially triggered by seepage forces in the test. It is obvious that soil nails cannot be used to stabilise a loose sand fill slope which has a high liquefaction potential.

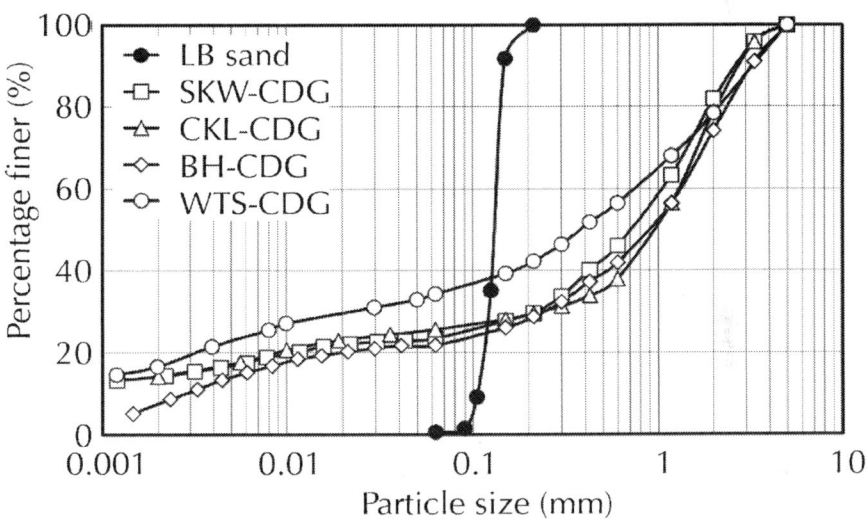

**Figure 18**: Particle size distributions of LB sand and CDG.

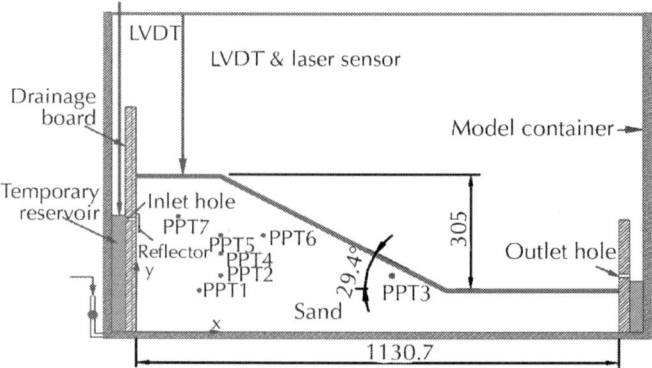

**Figure 19**: Centrifuge model of a loose sand fill slope subjected to rising ground water table at 60g(unit: mm) (Zhang and Ng, 2003).

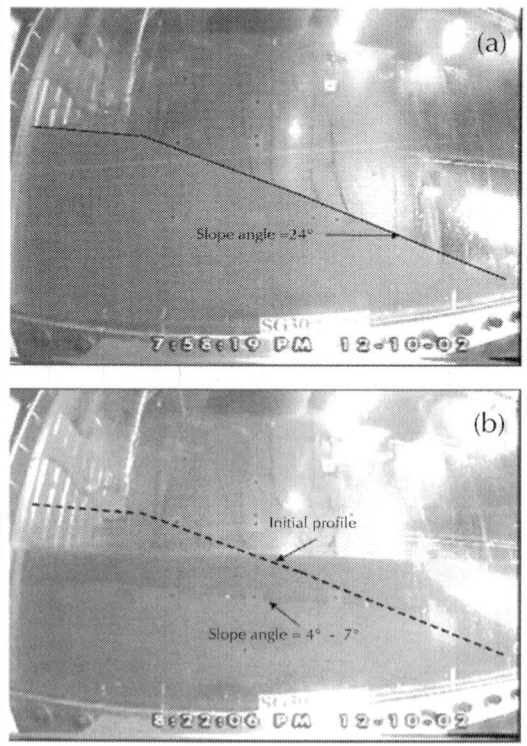

**Figure 20**: Slope profile in a loose sand fill test before rising ground water table (a) and after static liquefaction (b) (Zhang and Ng, 2003; Ng et al., 2007).

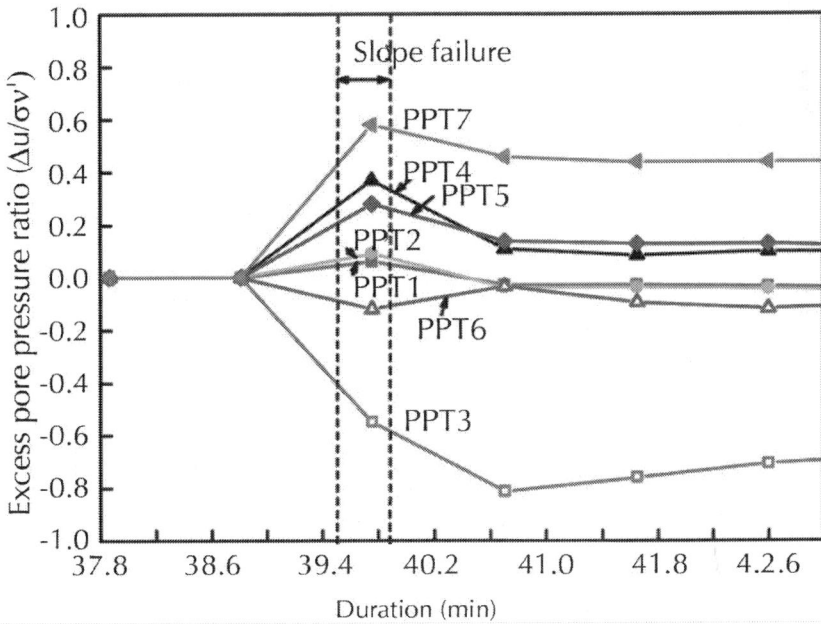

**Figure 21**: Measured sudden and substantial increases in excess pore water pressure at seven locations inside the slope (Zhang and Ng, 2003; Ng, 2009).

## *Non-Liquefied Slide of Shallow CDG Fill Slopes*

Apart from loose fill slopes using material with high liquefaction potential (i.e., LB), slopes loosely backfilled with a material with very small liquefaction potential (CDG) were also tested. Fig. 22 shows an instrumented centrifuge model for studying the potential static liquefaction of a loose shallow CDG fill slope subjected to a rising ground water table. The particle size distribution of the CDG used is denoted as WTS in Fig. 18. The initial fill density was 66%. This model was used to simulate a 1.5 m thick, 24 m high layered fill slope when tested at 60g. In addition to LSs installed for monitoring soil surface movements, PPTs were installed to measure excess pore water pressures during the tests. Effects of layering were considered by tilting the model container during model preparation. The slope was destabilised by downward seepage created by a hydraulic gradient, which was controlled by the water level inside the upstream temporary reservoir and the conditions of the outlet hole located downstream (Fig. 22). Two failures were induced in the test.

Figs. 23 and 24 show respectively the occurrence of a non-liquefied slide and the measured excess pore water pressure during two failures. The slide was initiated near the crest. Based on the observed failure mechanisms and the small excess pore water pressures measured, it was concluded that non-liquefied slide of loose shallow CDG fill slopes could occur but static liquefaction was very unlikely to happen. The significant difference between the observed centrifuge test results from the loose LB sand (Fig. 20b) and CDG fill slopes (Fig. 23) may be attributed to the difference in fine contents, gradation and liquefaction potential of the two materials.

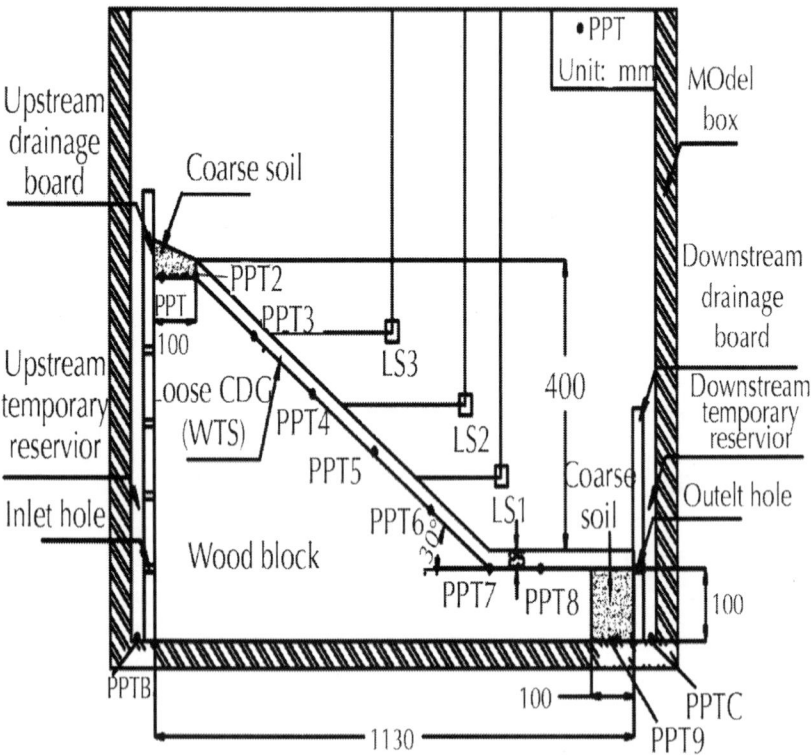

**Figure 22**: Model package of an instrumented shallow fill slope (Ng et al., 2007).

**Figure 23**: Top view of the model showing a non-liquefied slide (Ng et al., 2007).

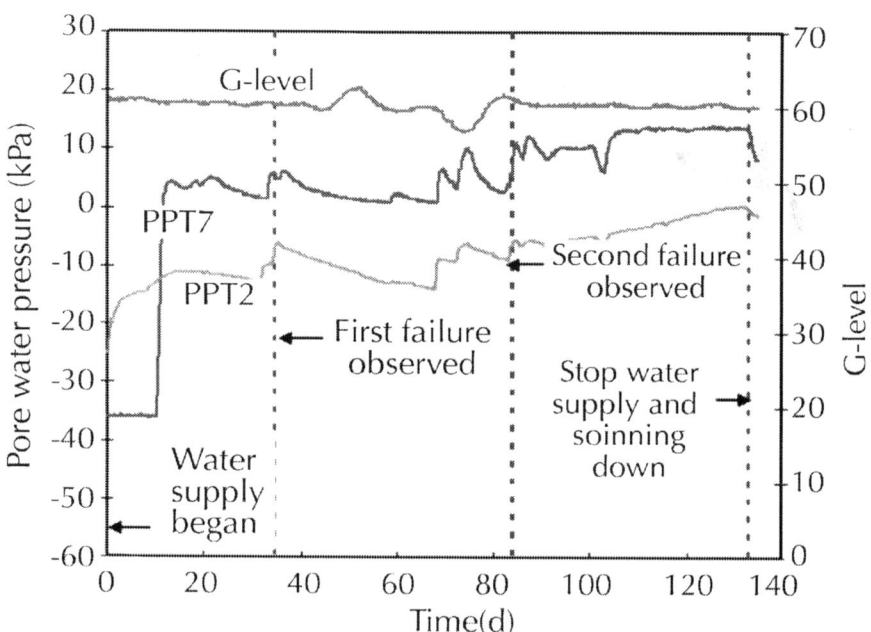

**Figure 24**: Variations in the measured pore water pressure at the crest (PPT2) and at the toe (PPT7) of the slope with time (Ng et al., 2007).

## *Response of Loose CDG Slopes to Earthquakes*

To further investigate the possibility of flow liquefaction of loose CDG fill slopes, uni-axial and bi-axial dynamic centrifuge tests were carried out (Ng et al., 2004b). The model CDG fill slopes were subjected to shaking ranging from 0.08g to 0.28g(prototype) in the centrifuge at HKUST. All the models were essentially the same in geometrical layout and made of loose CDG with the same initial dry density. Fig. 25 shows a typical model slope (6 m in prototype) initially inclined at 30° to the horizontal, with its instrumentation. A rigid rectangular model box was used to contain the CDG samples compacted to an initial dry density of about 1.4 g/cm$^3$ (or 77% of relative compaction). Five pairs of miniature accelerometers (ACCs) were installed in the slope. Each pair was arranged to measure soil accelerations in two horizontal directions (i.e., X- and Y-direction). Four miniature PPTs were installed in the soil near the accelerometers to record pore water pressures during shaking. On top of the slope, three LVDTs were mounted to measure the crest settlement, and one LVDT and one LS were used to measure horizontal movement of the crest.

To simulate the correct dissipation rate of excess pore pressures in the centrifuge tests, sodium carboxy methylcellulose (CMC) powder was mixed with distilled deionized water to form the properly scaled viscous pore fluid and to saturate the loose CDG model slopes.

After model preparation, the speed of the centrifuge was increased to 38g. Once a steady state pore pressure condition was reached at all transducers, a windowed 50 Hz (1.3 Hz prototype), 0.5 s (19 s prototype) duration sinusoidal waveform was then applied (Ng et al., 2004b). After triggering each earthquake, the centrifuge acceleration was maintained long enough to allow the dissipation of any excess pore pressure. This paper only highlights some important results from one biaxial shaking test. Other details of all the tests were presented in (Ng et al., 2004b).

Fig. 26 shows some measured horizontal acceleration time histories in the X- and Y-direction together with their normalised amplitudes in the Fourier domain in the bi-axial shaking test. The base input accelerations (recorded by ACC-T-X and ACC-T-Y as shown in the figure) were 11.26g (0.28g prototype) and 7.77g (0.19g prototype) in the X-direction and Y-direction, respectively. The windowed sinusoid waveform applied in the Y-direction lagged the X-direction input signal by 90°. Recorded by the accelerometer near the crest, the peak acceleration in the X-direction increased by 45% at ACC4-X, which is higher than that measured in a corresponding uni-axial shaking test (Ng et al., 2004b). A similar trend of variations in the acceleration was also found in the Y-direction. The normalised spectral amplitudes of acceleration at the predominant frequency of 50 Hz decreased by about 9% in the X-direction but increased by about 4% in the Y-direction in the upper portion of the embankment.

Fig. 27 shows the time history of the excess pore pressure ratios along the height of the model embankment during shaking. Peak acceleration occurred at about 0.25 s after the start of shaking. The maximum pore pressure ratio occurred at about 0.33 s at each of the three transducers (PPT1, PPT2, and PPT4). PPT1 and PPT2 recorded about the same maximum pore pressure ratio of 0.87, whereas PPT4 registered the smallest at 0.75. These measured values were less than the theoretical value of 1.0 for liquefaction, even though the pore fluid was correctly scaled in the test. The excess pore pressures dissipated to zero at about 12 s (6.8 min in prototype) after the start of shaking.

Fig. 28 shows a photograph of the model taken after the completion of a shaking test. The deformation profile for the slope was similar in both the uni-axial and bi-axial shaking tests. The observed profile of the deformed slope clearly illustrates that no liquefied flow and non-liquefied slide took place during the shaking. This suggests that loose CDG slopes are likely to be stable under the proposed design earthquake peak ground acceleration ranging from 0.08g to 0.11g in Hong Kong.

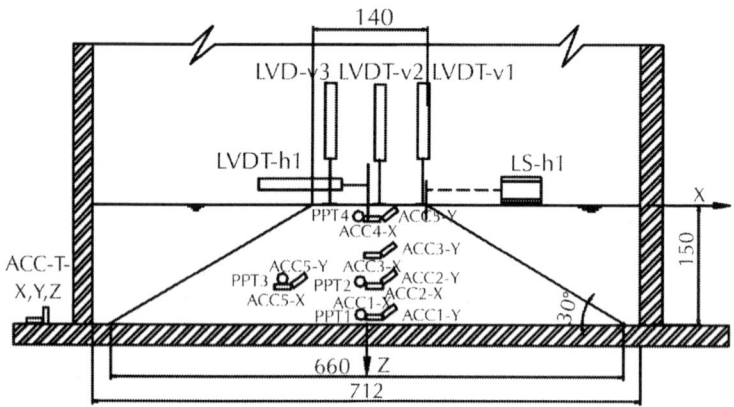

**Figure 25**: Configuration of the model slope and instrumentation (Ng et al., 2004b) (unit: mm).

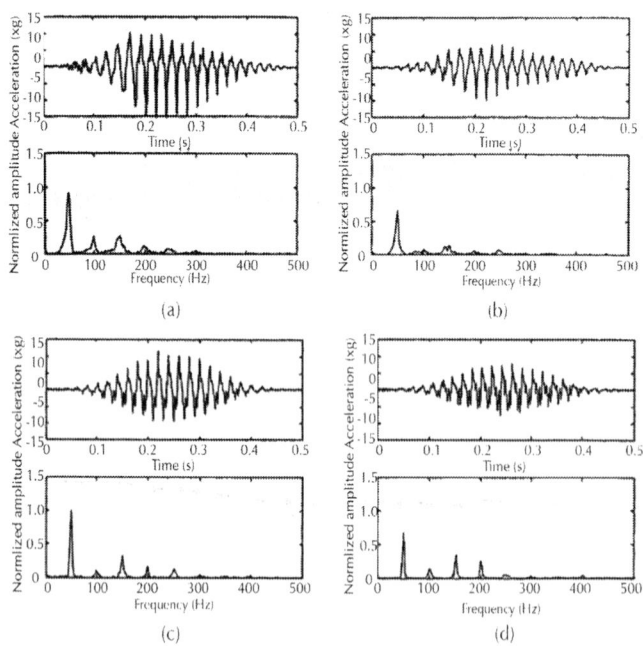

**Figure 26**: Seismic acceleration history and Fourier amplitude spectrum in the bi-axial shaking test (Ng et al., 2004b) (a) ACC4-X; (b) ACC4-Y; (c) ACC-T-X; (d) ACC-T-Y.

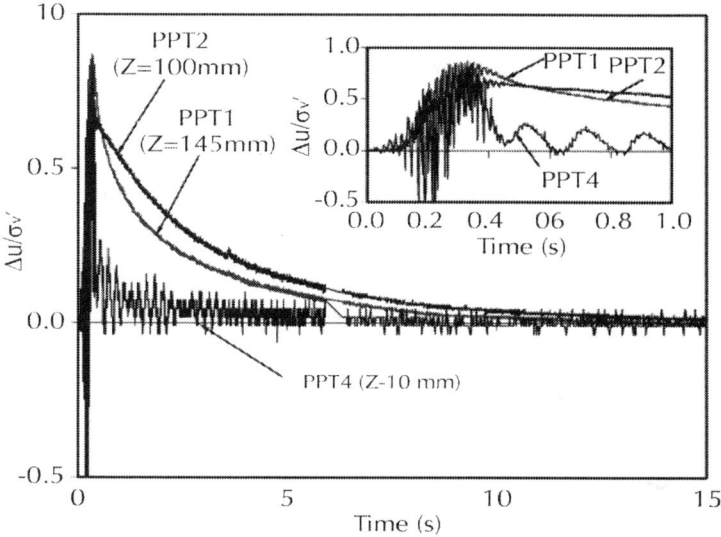

**Figure 27**: Measured excess pore water pressure ratios in the bi-axial shaking test M2D-0.3 (Ng et al.,2004b).

**Figure 28**: A typical profile of a loose fill slope after shaking (Ng et al., 2004b; 2007).

# Inter-Relationship between Centrifuge Modelling, Numerical Modelling, and Field Monitoring

Apart from centrifuge modelling, two other major tools used for solving geotechnical problems are numerical modelling and field monitoring. Fig. 29 illustrates the inter-relationship between centrifuge modelling, numerical modelling, and full-scale field monitoring/testing. These three approaches are complementary with each other since no approach is perfect for every geotechnical and geo-environmental problem in terms of quality and reliability of result, time, and cost. They each have their own advantages and disadvantages.

Although both centrifuge tests and full-scale field tests can provide physical data, many numerical modellers suggest using full-scale field tests and case histories to calibrate their constitutive models and model parameters. This type of calibration can be very misleading since compensating errors are often overlooked (Fig. 30). For instance, field data are always subjected to many uncertainties because the actual ground conditions, anisotropy in terms of strength, stiffness and permeability, degree of saturation, soil homogeneity and boundary conditions are normally not known for sure. Any computed results, which "match" with observed and measured field behaviours, may be fortuitous resulting from compensating errors.

On the other hand, a mismatch between computed and measured data does not necessarily imply that either field measurements or numerical predictions are incorrect. As illustrated in Fig. 31, there is a missing link in the procedure of calibrating numerical models and model parameters using field data. Prior to calibrating constitutive models and model parameters against field data/case histories, a vital intermediate step (i.e., physical model tests, either at 1g or high g) is desirable and necessary to provide 'known' boundary and ground conditions and soil parameters for numerical modellers since any physical model tests are man-made. Uncertainties in material properties, ground conditions, and boundary conditions can be minimised or even eliminated. Ideally, any numerical tool should be calibrated against measured data from well-controlled physical model tests first (such as flume and centrifuge tests), before trying to use it to predict actual field behaviour.

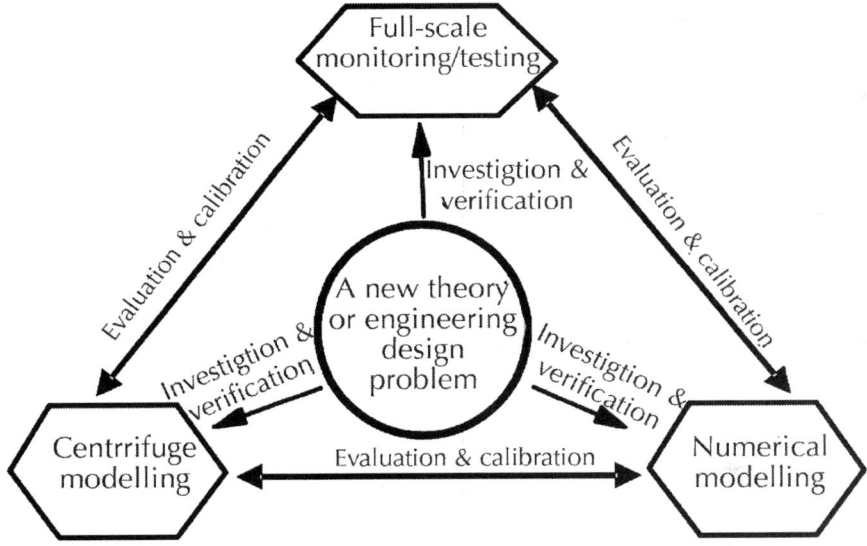

**Figure 29**: Inter-relationship between centrifuge modelling, numerical modeling, and full-scale field monitoring/testing.

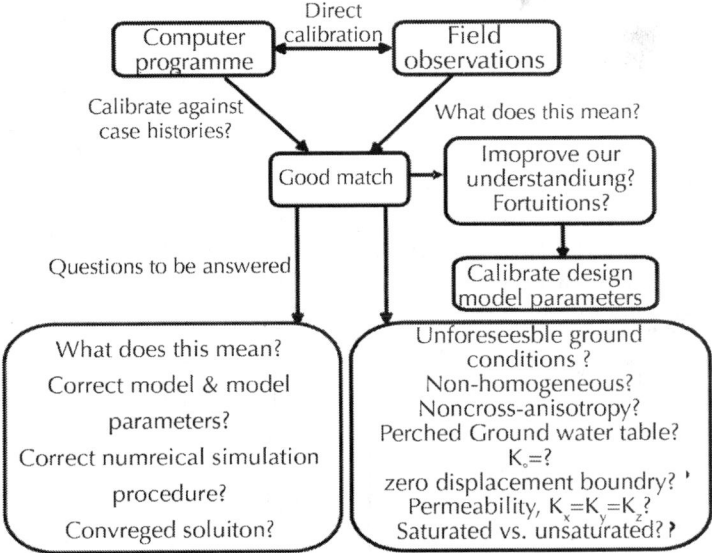

**Figure 30**: Concept of compensating errors.

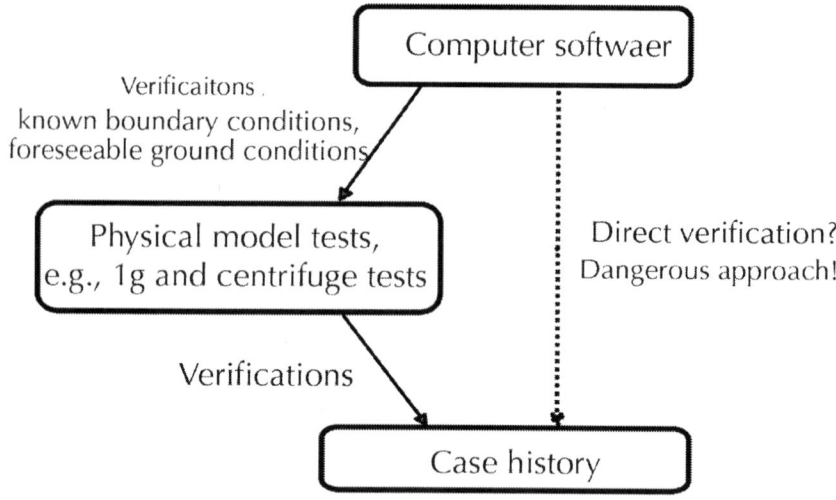

**Figure 31**: A missing link for calibrating numerical models and model parameters with field data/case histories.

# CONCLUSIONS

This ZENG Guo-xi Lecture reports four examples of the use of the state-of-the-art geotechnical centrifuge at the HKUST to investigate and understand complex geotechnical problems. Based on the centrifuge model tests, the following conclusions can be drawn:

Building tilt often results from non-homogeneity of the ground or from adjacent underground constructions. One possible method for correction of building tilt is to extract soil (i.e., creating stress release) from the less settled side of a building. To verify the effectiveness of this method, in-flight vertical boring adjacent to an initially tilted building was simulated in the centrifuge, by using the advanced four-axis robotic manipulator at the HKUST. The centrifuge test demonstrated that vertical soil extraction could effectively reduce building tilt. The test results also suggested that to properly predict the correction of building tilt by vertical soil extraction, construction sequences and the plastic behaviour of soil should be taken into account.

Three-dimensional centrifuge model tests were conducted to investigate the effect of tunnel collapse on an adjacent exsiting tunnel. The centrifuge test results showed that due to tunnel collapse, bending moments at the crown, invert, and the right springline (close to the collapse) of the adjacent exising tunnel were increased by 60%, 28%, and 228%, respectively.

Influence of excavation on the capacity of a single pile installed underneath was investigated. Centrifuge tests reveal that pile capacity after excavation can either increase or decrease, depending on roughness of the soil-pile interface. For low friction piles (piles in non-dilatant geo-materials), ultimate shaft resistance reduces in proportion to the vertical stress relief resulting from excavation. On the other hand, the capacity of high friction piles (piles in dilatant geo-materials) increases (by about 39%) after excavation. This is because reduced stress level in the soil after excavation makes the rough soil-pile interface more dilatant. The dilation at the soil-pile interface increases horizontal stress acting on the pile, and hence increases the ultimate shaft resistance.

To further clarify and better understand static liquifaction and its importance in slope engineering, triaxial element tests and centrifuge model tests on loose fill slopes using gap-graded LB sand and a well-graded silty sand (i.e., CDG) were carried out. In the centrifuge test, static liquefaction/fluidisation of the loose LB sand fill slope due to a rising ground water table was successfully simulated. In contrast, only non-liquefied slide was observed in loose CDG fill slopes when they were subjected to a rising ground water table. The distinct difference between the observed centrifuge tests on the LB sand and CDG fill slopes may be attributed to the differences in the fine contents, gradation, and liquefaction potential of the two materials.

Apart from static tests, the dynamic response of a loose CDG fill embankment was also investigated using the bi-axial shaking table in the centrifuge at the HKUST. The bi-axial shaking test with a peak horizontal base acceleration of about 0.3 resulted in maximum excess pore water pressure ratios ranging from 0.75 to 0.87. No evidence of flow liquefaction was observed, indicating that loose CDG slopes are likely to be stable under the proposed design earthquake peak ground acceleration ranging from 0.08 to 0.11 in Hong Kong. In addition to improving the understanding of complex geotechnical problems, well-controlled centrifuge tests can also provide high quality physical data to verify analytical and numerical methods. It is highly recommended that any numerical tool should be verified by well-controlled model tests (i.e., 1g or centrifuge tests) prior to predicting actual performce in the field, which often involves many more uncertainties than those in model tests. Although geotechnical centrifuge is a powerful physical modelling tool for researchers and engineers, it has some limitations. For instance, it is not suitable for use in investigating soil creep, ageing, and compensation grouting for tunneling.

# REFERENCES

1.    1990, Methods of Tests for Soils for Civil Engineering Purposes. British Standards Institution,London.

2.    Bucky,P.B, 1931, Use of models for the study of mining problems American Institution of Mining and Metallurgical Engineers, 425():3-28.

3.    Cai,Z.Y, 2001, A Comprehensive Study of State-dependent Dilatancy and Its Application in Shear Band Formation Analysis. PhD Thesis,The Hong Kong University of Science and Technology,Hong Kong, China.

4.    Castro,G, 1969,  Liquefaction of Sands. Harvard Soil Mechanics Series, No. 81, ():-.

5.    Cheney,J.A, 1988, . Centrifuges in Soil Mechanics,Balkema, Rotterdam.

6.    Craig,W.H, 1995, Geotechnical centrifuge: past, present and future  Geotechnical Centrifuge Technology, ():1-18.

7.    DeJong,J.T, Frost,J.D, 2002, Physical evidence of shear banding at granular-continuum interfaces , Proc. 15th ASCE Engineering Mechanics Conference, Columbia University, New York, NY, p. 8-.

8.    Fioravante,V, 2002, On the shaft friction modelling of non-displacement piles in sand  Soils and Foundations, 42(2):23-33.

9.    Garnier,J, Gaudin,C, Springman,S.M, 2007, Catalogue of scaling laws and similitude questions in geotechnical centrifuge modelling  International Journal of Physical Modelling in Geotechnics, 7(3):1-23.

10.   Hoek,E, 1965, The design of a centrifuge for the simulation of gravitational force fields in mine models Journal of South African Institute of Mining and Metallurgy, 65(9):455-487.

11.   2000, The collapse of NATM tunnels at heathrow airport. A Report on the Investigation by the Health and Safety Executive into the Collapse of New Austrian Tunnelling Method (NATM) Tunnels at the Central Terminal Area of Heathrow Airport on 20/21 October 1994. HSE Books,UK.

12.   Joseph,P.J, Einetein,H.H, Whiltman,R.V, 1988, A literature review of geotechnical centrifuge modelling with particular emphasis on rock mechanics. Massachusetts Institute of Technology,USA.

13.   Kimura,T, 1998, Development of geotechnical centrifuge in Japan  Proc, ():23-32.

14.   Kishida,H, Uesugi,M, 1987, Tests of the interface between sand and steel in the simple shear apparatus  Gotechnique, 37(1):45-52.

15.   Ko,H.Y, 1988, Summary of the state-of-the-art in centrifuge model testing  Centrifuge in Soil Mechanics, ():11-28.

16.   Kramer,S.T, 1996, Geotechnical Earthquake Engineering. Prentice Hall,New Jersey.

17.   Lei,G.H, Shi,J.Y, 2003, Physical meanings of kinematics in centrifuge modelling technique   Rock and Soil Mechanics, 24(2):188-193.

18.   Mindlin,R.D, 1936, Force at a point in the interior of a semi-infinite solid  Journal of Applied Physics, 7(5):195-202.

19.   Ng,C.W.W, 2005, Invited country report: failure mechanisms and stabilisation of loose fill slopes in Hong Kong  , Proceedings of International Seminar on Slope Disasters in Geomorphological/ Geotechnical Engineering, Osaka, p. 71-84.

20.   Ng,C.W.W, 2007, Liquefied flow and non-liquefied slide of loose fill slopes  , Proceedings of 13th Asian Regional Conference on Soil Mechanics and Geotechnical Engineering, Kolkata, Allied Publishers Private Ltd, p. -.

21.   Ng,C.W.W, 2008, Invited special lecture: deformation and failure mechanisms of loose and dense fill slopes with and without soil nails  , Proceedings of 10th International Symposium on Landslides and Engineered Slopes, Xian, China, p. 159-177.

22. Ng,C.W.W, 2009, What is static liquefaction failure of loose fill slope The 1st Italian Workshop on Landslides, Napoli, Italy, 1():91-102.

23. Ng,C.W.W, Xu,G.M, 2003, In-flight centrifuge modelling of vertical relief boring technique Chinese Journal of Geotechnical Engineering, 25(3):299-303.

24. Ng,C.W.W, Van Laak,P, Tang,W.H, 2001, The Hong Kong geotechnical centrifuge , Proceedings of 3rd International Conference on Soft Soil Engineering, Hong Kong, p. 225-230.

25. Ng,C.W.W, Yau,T.L.Y, Li,J.H.M, 2001, New failure load criterion for large diameter bored piles in weathered geomaterials Journal of Geotechnical and Geoenvironmental Engineering, 127(6):488-498.

26. Ng,C.W.W, Van Laak,P.A, Zhang,L.M, 2002, Development of a four-axis robotic manipulator for centrifuge modelling at HKUST , Proceedings of International Conference on Physical Modelling in Geotechnics, St. Johns Newfoundland, Canada, p. 71-76.

27. Ng,C.W.W, Zhou,X.W, Chung,J.K.H, 2003, Centrifuge modelling of multiple tunnel interaction in shallow ground , Proceedings of 13th European Conference on Soil Mechanics and Geotechnical Engineering, Prague, Czech Republic, p. 759-762.

28. Ng,C.W.W, Simons,N.E, Menzies,B.K, 2004, A Short Course in Soil-structure Engineering of Deep Foundations, Excavations and Tunnels. Thomas Telford,London.

29. Ng,C.W.W, Li,X.S, Van Laak,P.A, 2004, Centrifuge modelling of loose fill embankment subjected to uni-axial and bi-axial earthquakes Soil Dynamics and Earthquake Engineering, 24(4):305-318.

30. Ng,C.W.W, Pun,W.K, Kwok,S.S.K, 2007, Centrifuge modelling in engineering practice in Hong Kong , Geotechnical Division Annual Seminar, The Hong Kong Institution of Engineers, Hong Kong, p. 55-68.

31. Ng,C.W.W, Boonyarak,T, Man,D, 2013, Three-dimensional centrifuge and numerical modeling of the interaction between perpendicularly crossing tunnels Canadian Geotechnical Journal, 50(9):935-946.

32. Ng,C.W.W, Shi,J.W, Hong,Y, 2013, Three-dimensional centrifuge modelling of basement excavation effects on an existing tunnel in dry sand  Canadian Geotechnical Journal, 50(8):874-888.

33. Panek,L.A, 1949, Design of safe and economical structures Transactions of the American Institute of Mining and Metallurgical Engineers, 181():371-375.

34. Ramberg,H, 1968, Instability of layered systems in the field of gravity, I  Physics of the Earth and Planetary Interiors, 1(7):427-447.

35. Shen,C.K, Li,X.S, Ng,C.W.W, 1998, Development of a geotechnical centrifuge in Hong Kong  , Proceedings of Centrifuge, Tokyo, p. 13-18.

36. Taylor,R.N, 1995, Geotechnical Centrifuge Technology. Blackie Academic and Professional, London.

37. Terracina,F, 1962, Foundations of the Leaning Tower of Pisa Gotechnique, 12(4):336-339.

38. Zhang,M, 2006, Centrifuge Modelling of Potentially Liquefiable Loose Fill Slopes with and without Soil Nails. PhD Thesis, The Hong Kong University of Science and Technology, Hong Kong, China.

39. Zhang,M, Ng,C.W.W, 2003, Interim factual testing report I-SG30 & SR30. The Hong Kong University of Science and Technology, Hong Kong, China.

40. Zhang,M, Ng,C.W.W, Take,W.A, 2006, The role and mechanism of soil nails in liquefied loose sand fill slopes  , Proceedings of 6th International Conference Physical Modelling in Geotechnics, Hong Kong, p. 391-396.

41. Zheng,G, Peng,S.Y, Diao,Y, 2010, In-flight investigation of excavation effects on smooth single piles  , ():847-852.

42. Zheng,G, Peng,S.Y, Ng,C.W.W, 2012, Excavation effects on pile behaviour and capacity  Canadian Geotechnical Journal, 49(12):1347-1356.

# Chapter 9

# Uplifting Behavior of Shallow Buried Pipe in Liquefiable Soil by Dynamic Centrifuge Test

Bo Huang, [1] Jingwen Liu, [1] Peng Lin, [2] and Daosheng Ling[1]

[1]MOE Key Laboratory of Soft Soils and Geoenvironmental Engineering, Department of Civil Engineering, Zhejiang University, Hangzhou, 310058, China

[2]State Key Laboratory of Hydroscience and Engineering, Tsinghua University, Beijing 100084, China

## ABSTRACT

Underground pipelines are widely applied in the so-called lifeline engineerings. It shows according to seismic surveys that the damage from soil liquefaction to underground pipelines was the most serious, whose failures were mainly in the form of pipeline uplifting. In the present study, dynamic centrifuge model tests were conducted to study the uplifting behaviors of shallow-buried pipeline subjected to seismic vibration in liquefied sites. The uplifting mechanism was discussed through the responses of the pore water pressure and earth pressure around the pipeline. Additionally, the analysis of force, which the pipeline was subjected to before and during vibration, was introduced and proved to be reasonable by the comparison of the measured and the calculated results. The uplifting behavior of pipe is the combination effects of multiple forces, and is highly dependent on the excess pore pressure.

# INTRODUCTION

Pipelines are the artery of modern industries and urban life, widely used in lifeline engineerings such as water supply, electricity supply, natural gas transportation line, and communication cables. According to a large number of seismic disaster surveys [1, 2], the damage probability of the underground pipelines in liquefied soil is far greater than that in the nonliquefied soil. Shallow buried pipelines in liquefied sands might float upward displaying deflection deformation, and sometimes they even go above the ground, which often aggravates the damage degree. In the 1964's Niigata earthquake, among a total pipeline length of 470 kilometers, 68% of pipelines were destroyed. The pipelines located under water were damaged particularly seriously, whose failures were mainly due to uplifting deformation. Hereafter, "uplifting" phenomena of underground pipelines and other underground structures were more and more frequently observed in seismic incidents, such as the Loma Prieta earthquake and the Nansei-Oki earthquake [3–6].

There have been quite a lot of studies concentrating on the "uplifting" phenomenon of pipelines during soil liquefaction processes. The factors that affect the stress and deformation of pipelines during floating; for example, the buried depth, diameter, thickness, stiffness of the pipelines, the type, liquefied area, stiffness, and strength of soil, have been studied [7, 8]. It is generally considered that the diameter of pipeline, liquefied depth, stiffness of soil are critical parameters which have great effects on the structure deformation of buried pipeline; the largest floating displacement occurs in the central of the pipe. Some experts made a number of numerical and experimental researches on the response differences between the free field and field with pipelines during seismic vibration. Kitaura et al. [9] developed a hybrid procedure to study the pipeline response in the liquefied field. The numerical model shows that the pipeline response to seismic vibration is more significant when the excess pore pressure is low, and the buoyancy force increases with increasing excess pore pressure. Ling et al. [10] conducted centrifuge tests to investigate the seismic response differences of free field and field with pipelines with respect to the acceleration, excess pore pressure and settlement of ground surface, and so forth. Other experts [11, 12] investigated the effects of dilatancy angle and relative density of soil, diameter and buried depth of pipeline, underground water table level and thickness of the saturated soil layer, and so forth, on the uplifting behavior of pipeline, which indicated that the buried depth of pipe had the most significant impact.

Despite that the above mentioned studies have investigated the pipeline uplifting behaviors in liquefied fields with respect to influencing factors involving the pipe itself and the soil properties, through a comprehensive means of field investigations, numerical simulations, and model tests, there still exists disagreement on the understanding of the mechanism of pipeline uplifting. Some [13, 14] believe that the uplifting of pipelines is associated with the loss of soil shear strength due to soil liquefaction. Others [15] come to a conclusion that uplifting is simply related to the vibration rather than soil liquefaction, based on the observed phenomenon that uplifting starts when the soil is not fully liquefied, and ceases when the shaking is finished even when the excess pore pressure is still very high. Due to lack of understanding of the pipeline uplifting mechanism, different approaches are adopted to calculate the buoyancy force that pipelines are subjected to during soil liquefaction. In general, it is calculated using the formula, where the saturated soil is considered as fluidized material having a unit weight equivalent to its saturated unit weight [16–18]. It is worth mentioning that the buoyancy force is estimated in terms of excess pore pressure as well [19].

It is significantly favorable that researches are made on the stress conditions of the pipelines during uplifting for an in-depth understanding of the mechanism of pipe uplifting behaviors. In the present study, dynamic centrifuge model tests were conducted to investigate the mechanism of pipeline uplifting phenomenon during soil liquefaction.

Based on the measurements of acceleration, excess pore pressure and earth pressure around the pipelines, the forces on the pipelines before and during soil liquefaction were estimated, and the mechanism and the main influencing factors of pipeline uplifting were analyzed.

# TEST EQUIPMENT AND PROGRAMS

## Centrifuge, Shaking Table, And Rigid Container

The tests are conducted on the ZJU400 centrifuge with a shaking table, shown in Figure 1. The beam type centrifuge, with a payload capacity of 400 gt, has double platforms and an effective arm radius of 4.5 m. The maximum centrifugal acceleration is 100 g for dynamic tests. The centrifuge platforms have an overall dimension of 1.5 m × 1.2 m × 1.5 m. Meanwhile, an in-flight uniaxial electrohydraulic shaking table has been made to simulate seismic excitation. The shaking table has vibration frequencies ranging from 10 Hz~200 Hz. Its payload capacity is 500 kg, and its maximum lateral displacement and acceleration are 0.6 cm and 40 g, respectively. More details about the device can be found in [20].

**Figure 1**: Centrifuge and shaking table.

A Rigid container was used to prepare the model, whose inner dimension is 0.6 m (length) × 0.4 m (width) × 0.5 m (height), and its front perspex made window is convenient for direct observation of the experimental phenomena. A 25 mm thick piece of mouldable Duxseal was placed on each side of the container to reduce reflecting incident stress waves by at least 65% [21].

# Model Pipe

The model pipes are made of aluminum tube, with a density of 2.7 g/cm³, a length of 390 mm, and an inner and outer diameter of 36 mm and 40 mm, respectively. And they are used to simulate large diameter pipes like oil or gas pipelines. Each end of the pipes was sealed by a perspex disc with PTFE, and petroleum jelly was also used to reduce end friction. Microearth pressure transducers were installed on the bottom, side and crown of pipes to measure earth pressures (the normal stress).

Two model pipes were buried in the ground. One was used for measuring the uplift displacement, named pipe 1#; the other was used for measuring the stabilizing force during soil liquefaction, named pipe 2#. Pipe 1# could move freely during tests. Pipe 2# was installed to the rigid container through a connecting rod. The force, provided by the connecting rod to keep pipe 2# stable in the vertical direction, was defined as stabilizing force. In order to acquire the stabilizing force of pipe in the centrifuge, a load cell was installed on the connecting rod.

In order to measure vertical displacements of underground structures in the centrifuge, draw-wire displacement potentiometers are commonly used [22]. However, the potentiometer cable has tension force which will reduce the structures self-weight to some extent. Moreover, the tension force varies with the centrifugal acceleration, which is hard to be calibrated. Therefore, two aluminum alloy spokes with discs on the end were installed on pipe 1#, as shown in Figure 2. The vertical displacements of the discs could be measured by potentiometers, which guaranteed a more precise and reliable measurement while pipe 1# moved freely.

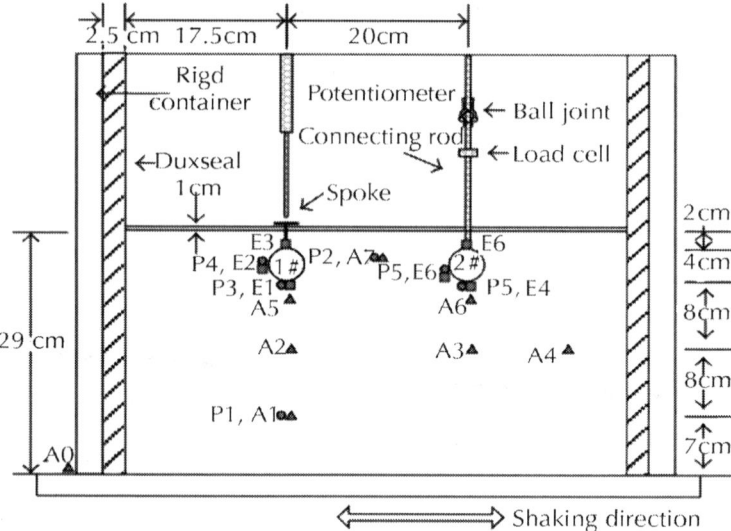

**Figure 2:** The layout of pipes and sensors.

A simple device of a connecting rod with ball joint was developed, as shown in Figure 2. This device kept pipe 2# stable in the vertical direction, and meanwhile moved freely in the horizontal direction. As pipe 2# was stable in the vertical direction, the shear strength of the overlaying soil would not take a part in the measured data of the force that pipe 2# is subjected to during vibration.

# Sand and Viscous Fluid

Fujian standard sand, which is widely used in China for geotechnical physical modeling tests [23, 24], was adopted in the present study. It has a mean diameter ($D_{50}$) 0.16 mm; the uneven coefficient ($C_u$) and the curvature coefficient ($C_c$) are 1.6 and 0.95, respectively. The maximum and the minimum void ratios are 0.96 and 0.61, respectively. The model foundation was prepared by pluviation method. The sand was rained from a sieve in a hopper into the container, where the falling height of the hopper and the shape of the sieve were kept unchanged based on the precalibrated results to obtain a constant relative density. The designed relative density of the two tests was 60%. The heights of the model foundation were 29 cm and 32 cm in test 1 and test 2, respectively.

There is a conflict between dynamic and permeability time scale, for the former is $1/n$, and the latter is $1/n^2$. To solve the problem, viscous fluid was introduced to reduce the permeability of soil. Methyl cellulose fluid, which is commonly used in geotechnical centrifuge modeling tests, has similar compressibility and density to water [25]. Furthermore, it is also capable of sustaining high pore pressure for liquefaction studies and can have any viscosity by changing the mixture ratio of hydroxypropyl methylcellulose (HPMC) powder to water. Based on the precalibrated relationship of mixture ratio, permeability, and temperature, the mixture ratio was determined to be 3.0% at a centrifugal acceleration 30 g. Methyl cellulose fluid was prepared in water with a temperature of 70°C and introduced at a rate of 0.1 L/h into the model foundation when cooling down, which was slow enough to avoid sand boil phenomenon. The vacuum method is chosen to saturate the soil, the whole process of saturation took over 200 h and the water level was kept 1 cm above the ground when the saturation is completed.

# Seismic Excitations

Three types of excitation waves were adopted, that is, EL-Centro wave, Taft wave, and Zhejiang seism wave. El-Centro wave was recorded in the Imperial Valley earthquake of California in 1940, with a primary period of 0.5 s, belonging to near earthquake. Taft wave was recorded in the earthquake happened in Kern of California in 1952, with a primary period of 0.5 s, belonging to distant earthquake. These two waves are commonly used. Zhejiang seism wave is an artificial seismic wave suited the seismic zoning type of Zhejiang Province in China, with a 10 s duration. Assuming the exceeding probability of Zhejiang seism wave to be 10% and 2%, the maximum acceleration is 63.1 cm/$s^2$ and 153 cm/$s^2$, respectively.

# Testing Procedures

In the present study, two centrifuge tests were conducted on pipes of different buried depths under the same ground conditions. The centrifuge accelerations were both 30 g. The buried depths, measured from the top of pipes to the ground surface, were 20 mm (equal to 0.5D) for test 1 and 80 mm (equal to 2D) for test 2. Accelerometers, pore pressure transducers, earth pressure transducers, potentiometers, and load cell were used in the tests [26]. The layout of the sensors and pipes for each test is shown in Figure 2.

When started, the centrifuge was accelerated to 30 g gradually. The relative densities of the ground before shaking were 65.2% for test 1 and 61.9% for test 2. The excitation progress was divided into 3 stages based on the acceleration amplitudes from weak to strong. White noise excitations were applied before and after each stage to test the dynamic characteristics of the model. The schedule of excitations as well as the uplifting status of the pipe 1# at each shaking stage is shown in Table1. There was at least a 30 min interval between two shaking stages, so that the excess pore pressure can dissipate entirely. After all the excitations were applied, the test data as will be mentioned in the following sections are converted to the prototype scale.

**Table 1:** Centrifuge testing program and uplifting status of pipe 1# during tests

| Test number | Seismic excitation | | | |
|---|---|---|---|---|
| | Seismic wave | Duration (s) | Amplitude (g) | Uplifting status |
| Test 1 | Noise | 30 | 0.02 | |
| | Zhejiang seism wave | | 0.1 | Remain still |
| | EL-Centro | | 0.1 | Remain still |
| | Noise | | 0.02 | |
| | Zhejiang seism wave | | 0.15 | Sink slightly |
| | EL-Centro | | 0.15 | Rise slightly |
| | Noise | | 0.02 | |
| | EL-Centro | | 0.5 | Rise |
| | Taft | | 0.4 | Rise |
| | Noise | | 0.02 | |

| Test 2 | Noise | 30 | 0.02 | |
|---|---|---|---|---|
| | EL-Centro | | 0.1 | Remain still |
| | Zhejiang seism wave | | 0.1 | Remain still |
| | Noise | | 0.02 | |
| | EL-Centro | | 0.4 | Rise |
| | Taft | | 0.4 | Rise |
| | Noise | | 0.02 | |

# RESULTS OF TESTS

## The Degrees of Soil Liquefaction and the Pore Pressure Response around the Pipe

### *The Degrees of Soil Liquefaction*

The excess pore pressure ratio $\Delta u / \sigma'$ defined as the value of the excess pore pressure normalized by the initial vertical effective stress represents the degree of soil liquefaction. If the value of $\Delta u / \sigma'$ reaches one, it means the soil is fully liquefied. As the build-up of excess pore pressure in the two tests was similar, only part of the results is given, that is, the variations of $\Delta u / \sigma'$ in test 2 under 0.1 g and 0.4 g excited by El-Centro wave, as shown in Figure 3. The buried depths of P1 and P5 were 8.4 m and 3.6 m, respectively. It shows that the excess pore pressure generated from the start the excess pore pressure ratio defined as the value of the excess pore pressure normalized by the initial vertical effective stress represents the degree of soil liquefaction. If the value of reaches one, it means the soil is fully liquefied. As the build-up of excess pore pressure in the two tests was similar, only part of the results is given, that is, the variations of in test 2 under 0.1 g and 0.4 g excited by El-Centro wave, as shown in Figure 3. The buried depths of P1 and P5 were 8.4 m and 3.6 m, respectively. It shows that the excess pore pressure generated from the start of the earthquake vibration. As soon as the vibration stopped, the excess pore pressure ratio began to dissipate. In the two tests, although no excitations led the soil to fully liquefied state, the "uplifting" phenomenon still existed, which suggests that there is a high potential for the occurrence of pipe uplifting in incompletely liquefied soil the earthquake vibration. As soon as the vibration stopped, the excess pore pressure ratio began to dissipate. In the two tests, although no excitations led the soil to fully liquefied state, the "uplifting" phenomenon still existed, which suggests that there is a high potential for the occurrence of pipe uplifting in incompletely liquefied soil.

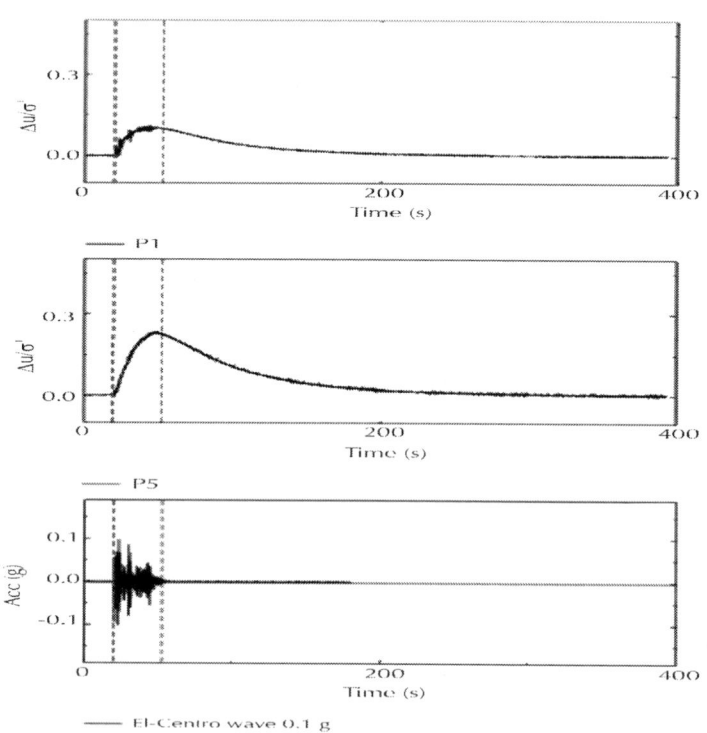

(a)

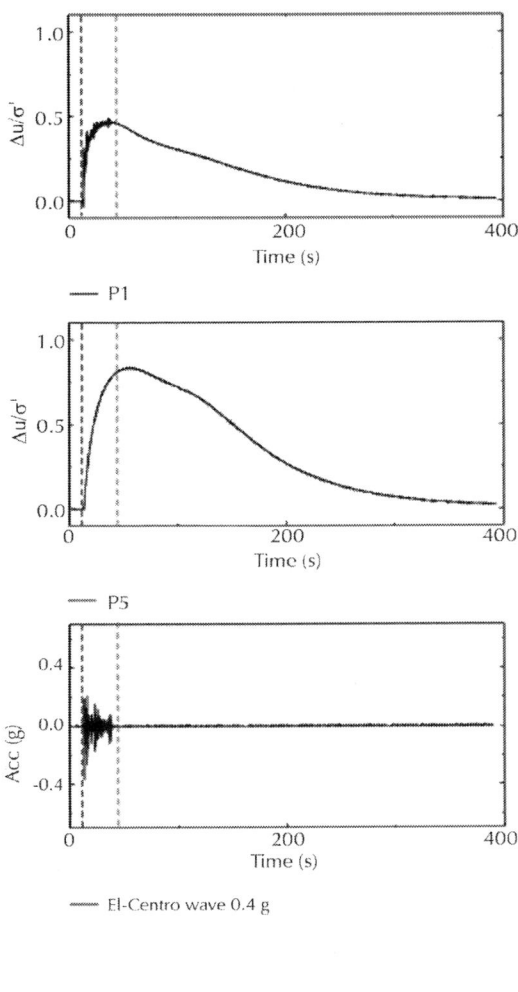

(b)

**Figure 3:** Ground response of $\Delta u / \sigma'$ in test 2 under El-Centro wave. (a) $a_{max}$ = 0.1g, (b) $a_{max}$ = 0.4g.

## The Pore Pressure Response around the Pipe

The variations of $\Delta u / \sigma'$ around the pipe in test 1 under 0.15 g and 0.5 g excited by El-Centro wave are shown in Figure 4. P3 and P4 were fixed at the bottom and side of pipe 1#, 1.8 m and 1.2 m below the surface, respectively. As the buried depth of pipes in test 1 was so shallow that the pore pressure could not be measured well at the crown of pipe; therefore, no transducer was installed there. In test 2, transducers P2, P3, and P4 were installed at the bottom, side, and crown of pipe 1#, respectively. P6 was embedded in the soil layer overlying pipe 1#. The variations of $\Delta u / \sigma'$ for P3, P4, and P6 under 0.1 g and 0.4 g excited by El-Centro wave are given in Figure 5. Due to the damage of P2, there is no measured data from P2.

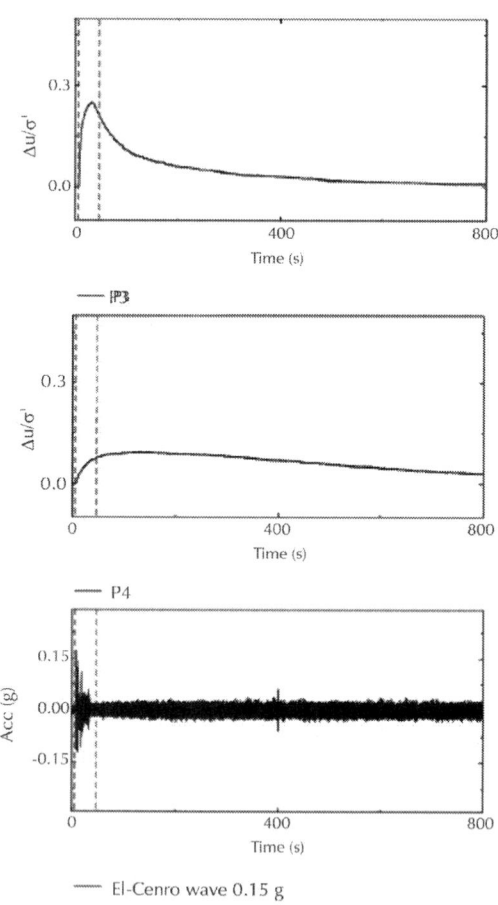

(a)

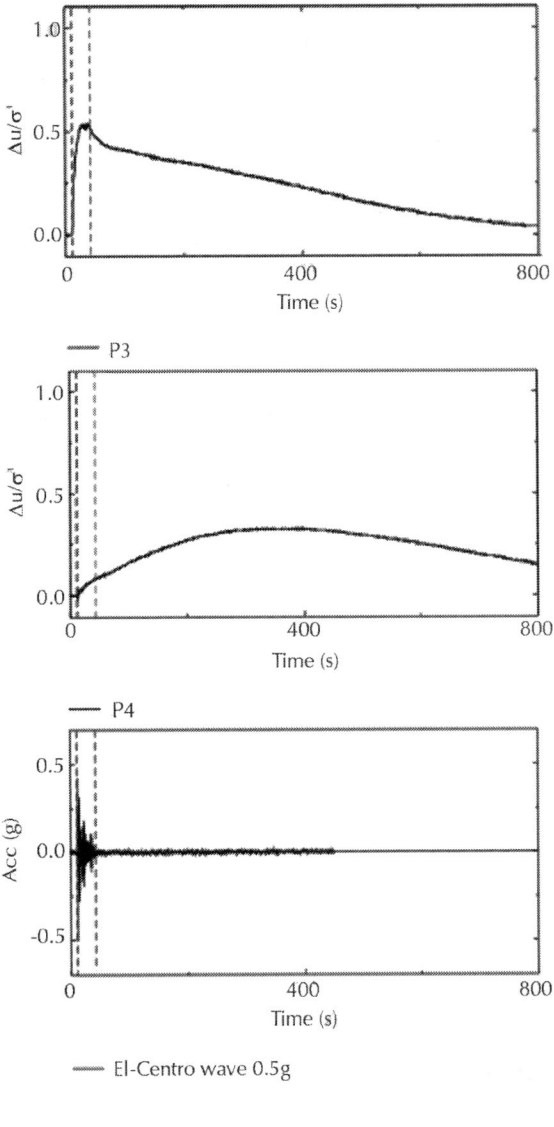

(b)

**Figure 4:** Pipe response of $\Delta u / \sigma'$ in test 1 under El-Centro wave. (a) $a_{max} = 0.15g$, (b) $a_{max} = 0.5g$.

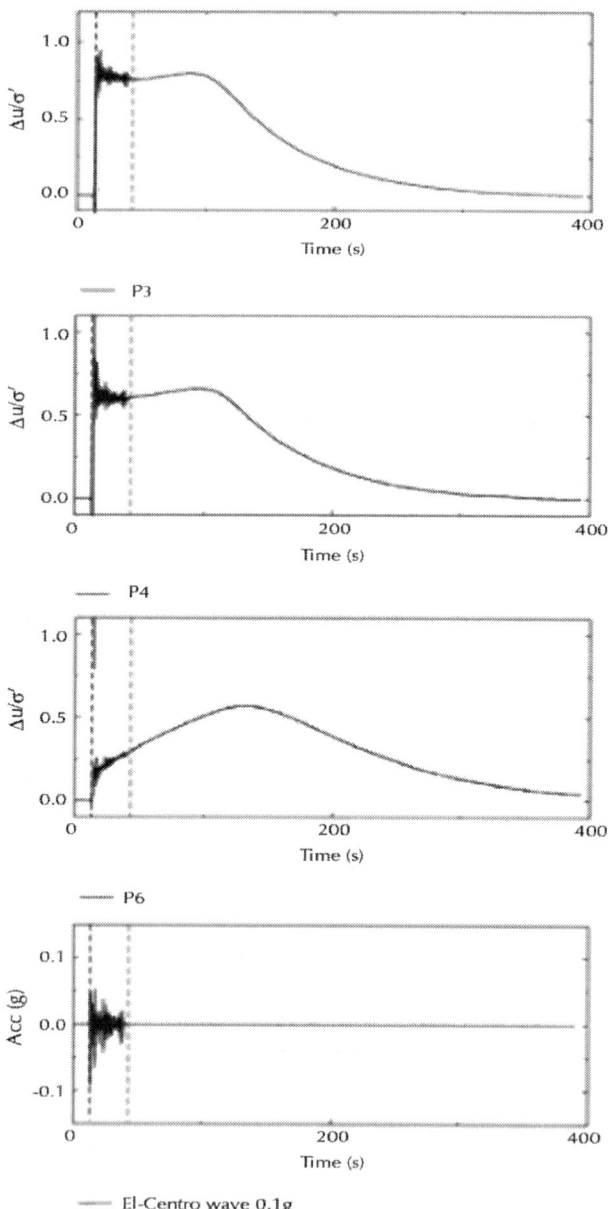

(a)

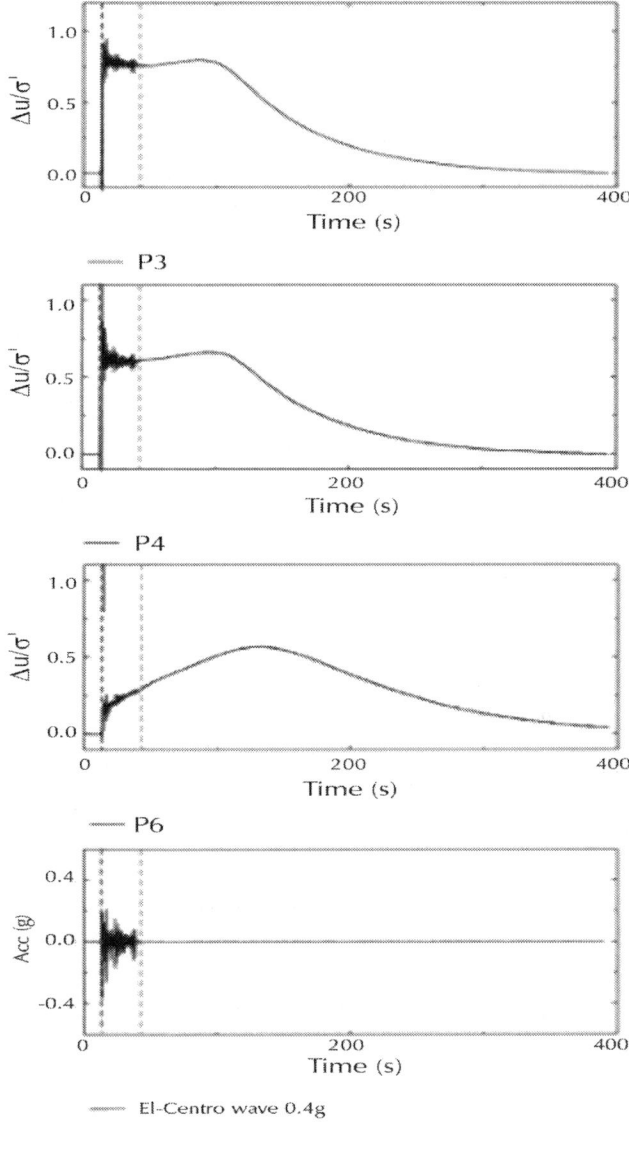

(b)

**Figure 5:** Pipe response of $\Delta u / \sigma'$ in test 2 under El-Centro wave. (a) $a_{max} = 0.1g$, (b) $a_{max} = 0.4g$.

It can be figured out from Figures 4 and 5 that excess pore pressures around the pipes respond rapidly once the earthquake load is applied. For a small amplitude excitation, the excess pore pressure dissipates gradually when the excitation ends. Considering that the amplitude of real earthquake wave decreases obviously at the late stages, the excess pore pressures might even dissipate before the end of vibration, as is shown in Figure 5(a). For stronger seismic excitations, the excess pore pressures of soil below the pipe dissipate when the vibration is ceased, but the dissipation rate slows down obviously, as shown in Figure4(b). Meanwhile, at the side and crown of pipe, the excess pore pressures retains for a while after the vibration is ceased. This phenomenon is attributable to the supply of the pore fluid draining from the base of the rigid container, which was more sufficient than the dissipation of the excess pore pressures at shallow location, so that the excess pore pressures at shallow places keep generating, as shown in Figure 4(b) (location P4) and Figure 5(b) (location P6).

It can be seen that the stronger the excitation is, the larger the excess pore pressures ratio will be the dissipation rate of the excess pore pressure decreases with the decreasing buried depth of the pipe. And in some cases, the excess pore pressures even keep generating The frictional resistance between soil grains was largely reduced by the increase in pore pressure Therefore, pipe floats upward more easily through the soil on the condition of stronger excitations or lower buried depths of pipe.

# Earth Pressure Response

The layout of the earth pressure transducers is shown in Figure 2. And the earth pressure (the total stress, which contains both effective stress and pore pressure) around the circumference of pipe under 0.15 g and 0.5 g excited by El-Centro wave in test1 is shown in Figure 6. The value of earth pressure changed while the vibration was activated and recovered gradually to 0 after the vibration ended. The responses of the earth pressure were different between pipe 1# and pipe 2# due to the different constraint conditions. Responses of pipe 1# which could move freely were larger than that of pipe 2# which was fixed in the vertical direction. It can be seen in Figure 6 that the increments of the earth pressure at the bottom and side of pipe are proportional to the amplitude of the excitations. However, response of the earth pressure at the crown was weak or even showed a slight decrease, which might indicate the weight reduction of the overlay soil.

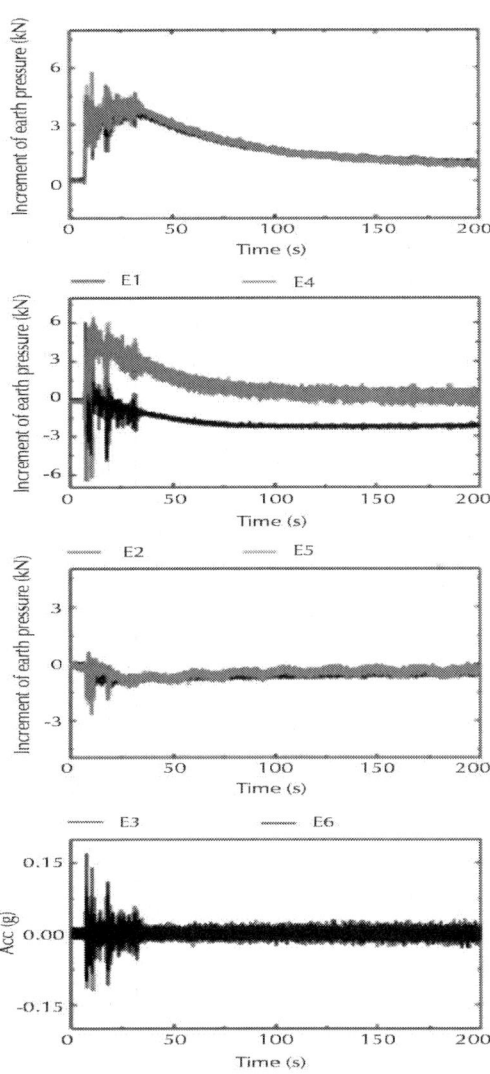

(a)

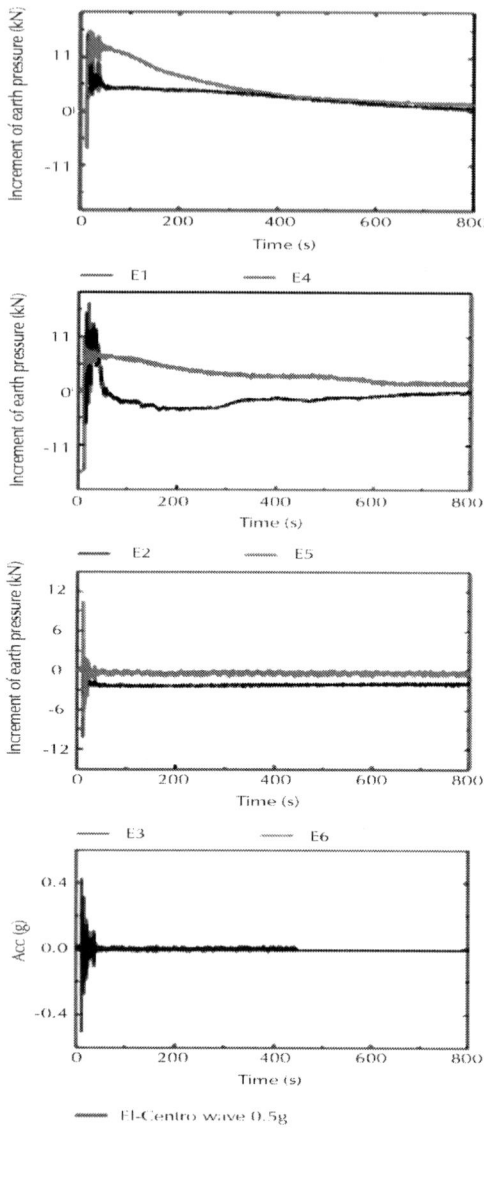

(b)

**Figure 6**: Increment of earth pressure in test 1 under El-Centro wave. (a) $a_{max}$ = 0.15 g and (b) $a_{max}$ = 0.5 g.

## Pipe Uplifting

It is found that pipeline uplifting takes place once the vibration starts and ceases when the vibration stops despite the presence of high excess pore pressures. Some researchers hold the view that the uplifting of the pipe is highly dependent on the input earthquake motion and weakly related to the increase of excess pore water pressure [15, 19]. Figure 7 shows uplifting responses of pipe 1# under Taft wave in test 1 and test 2. And uplifting responses of pipe 1# under different amplitudes of EL-Centro wave are shown in Figure 8.

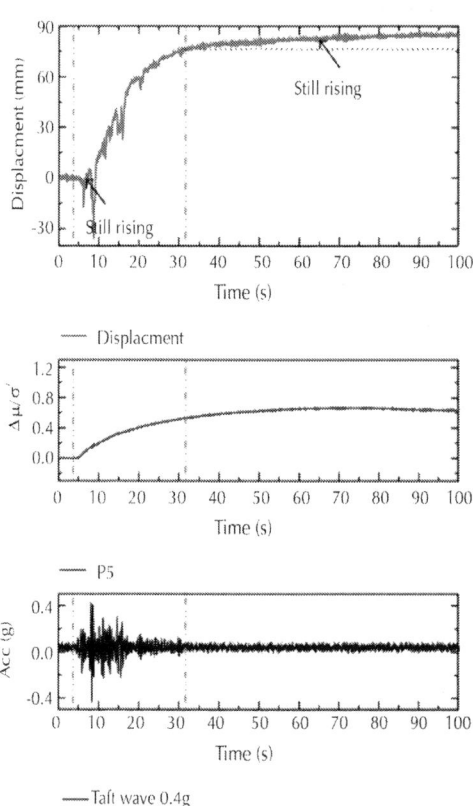

(a)

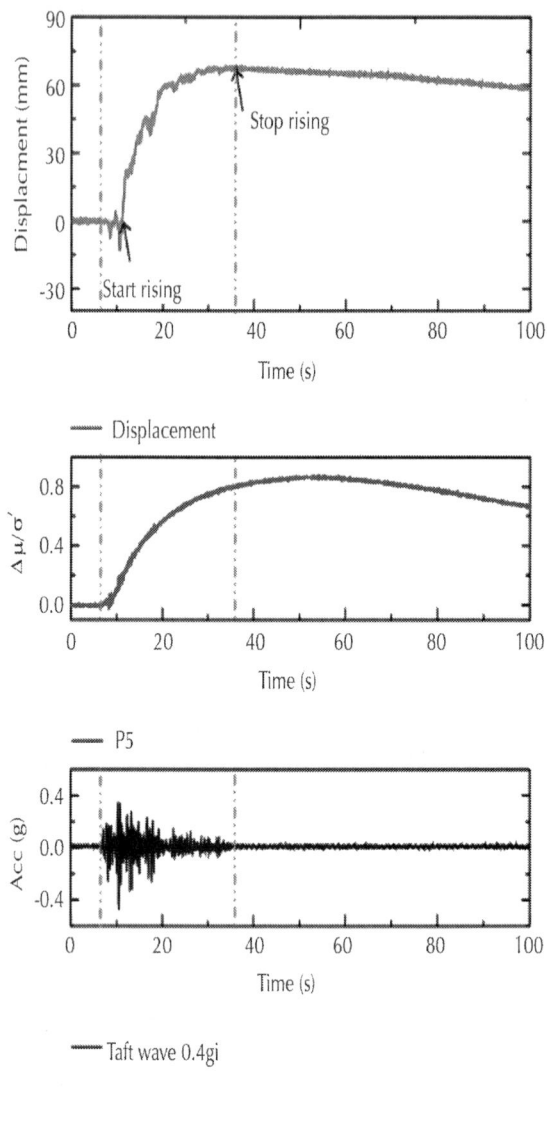

(b)

**Figure 7**: Uplift displacement of pipe 1# at different depth under Taft wave.
(a) H/D = 0.5 and (b) H/D = 2.

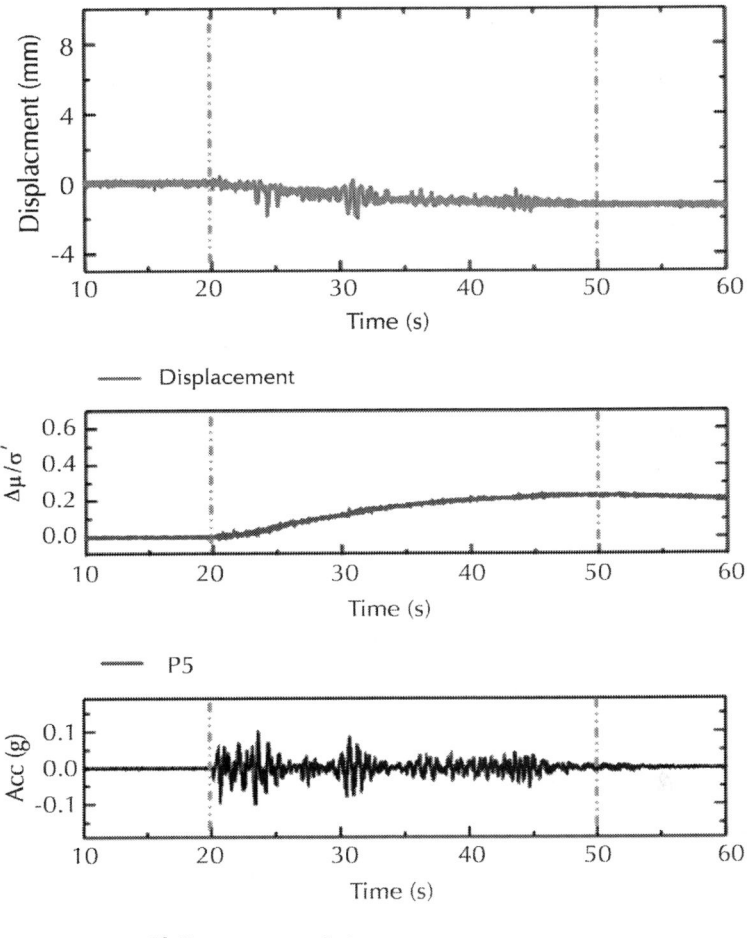

El-Centro wave 0.1g

(a)

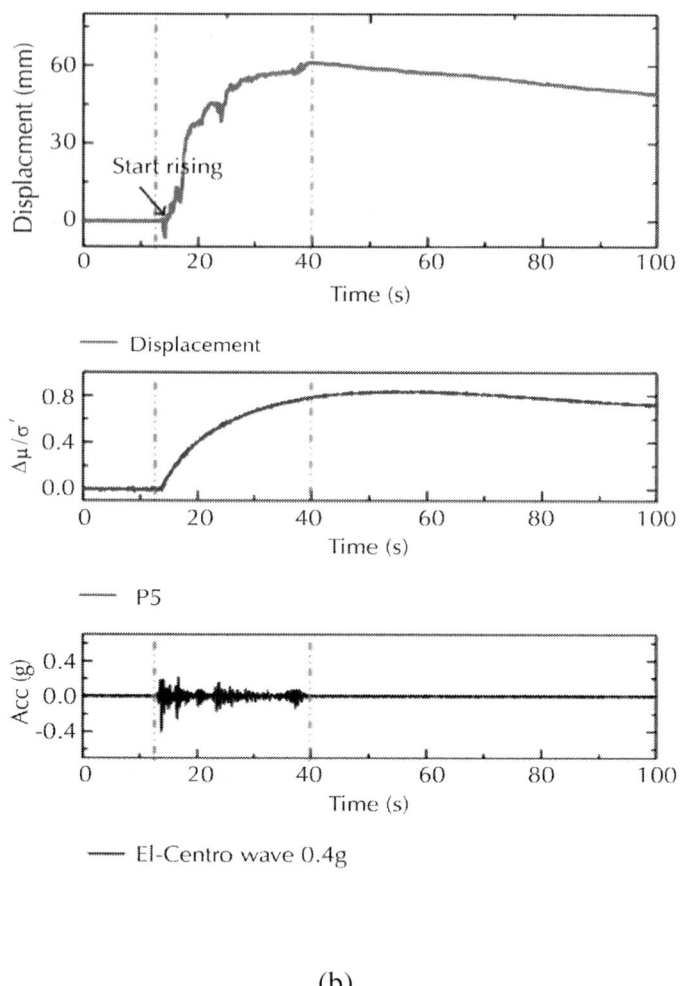

(b)

**Figure 8**: Uplift displacement of pipe 1# in test 2 under El-Centro wave. (a) $a_{max}$ = 0.1 g and (b) $a_{max}$ = 0.4 g.

It is seemingly that the uplifting phenomenon of pipe occurred after shaking and ceased when the shaking ceases. Nevertheless, the uplifting movement is not directly determined by the shaking itself but the response of the pore pressure and the soil pressure. It can be seen that the uplifting of the pipe takes place only after considerable excess pore pressure is generated, rather than immediately after the vibration started. The excess pore pressure ratios of P5 (which were at the same depth of the bottom of pipe 1#) when pipe 1# began to move up were shown in Figure 9. The excess pore pressure ratios distributed between 0.1 and 0.2. However, it does not mean that the pipe will float upward once the excess pore pressure ratio has reached 0.2. As can be seen in Figure 8(a), the pipe settled along with the soil particles under 0.1 g excited by EL-Centro wave, although the maximum excess pore pressure ratio was above 0.2.

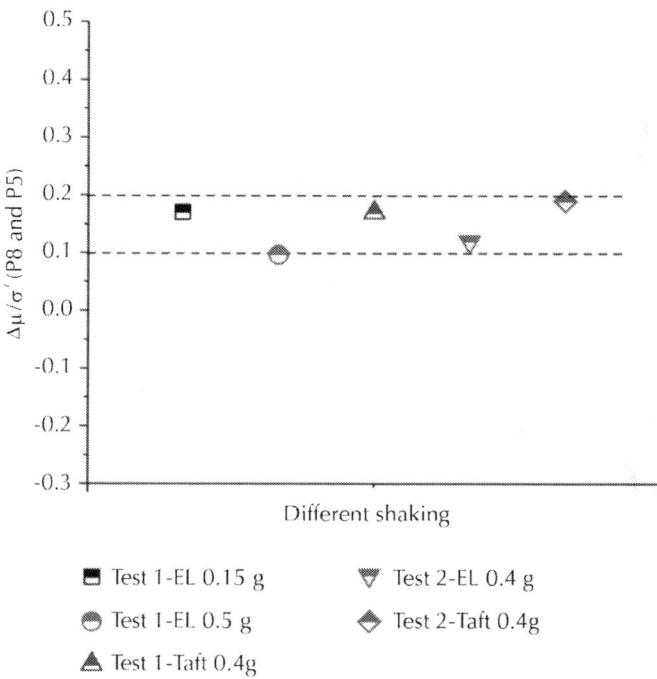

**Figure 9**: Excess pore pressure of P5 when pipe 1# starts to float.

Actually, the maximum uplift displacement was not present right after the vibration stopped each time. It can be seen in Figure 7(a), the tendency of uplifting still existed when the earthquake ceased. The uplifting behavior of pipe in liquefied soil is a multiforce coupled behavior, which is not only dependent on the build-up of excess pore pressure but also determined by the shear strength of the soil, relative displacement of pipe and soil, the amplitude of the input seismic wave, and so forth.

# The Response of Stabilizing Force

The stabilizing force, which kept the pipe 2# stable in the vertical direction, was measured by a load cell fixed on the device. The variation of stabilizing force in test 1 under El-Centro wave with different amplitudes is shown in Figure 10. Herein, negative values of the stabilizing force mean that the pipe is prone to settle down, as seen in Figure 10(a), and positive values represents that the pipe has the tendency to uplift, as shown in Figures 10(b) and 10(c).

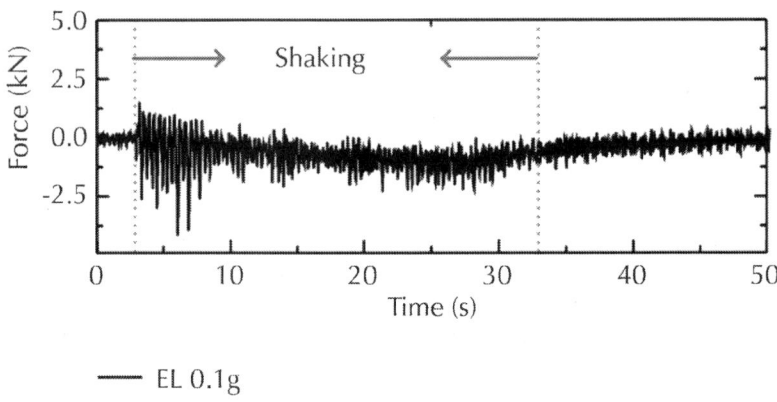

EL 0.1g

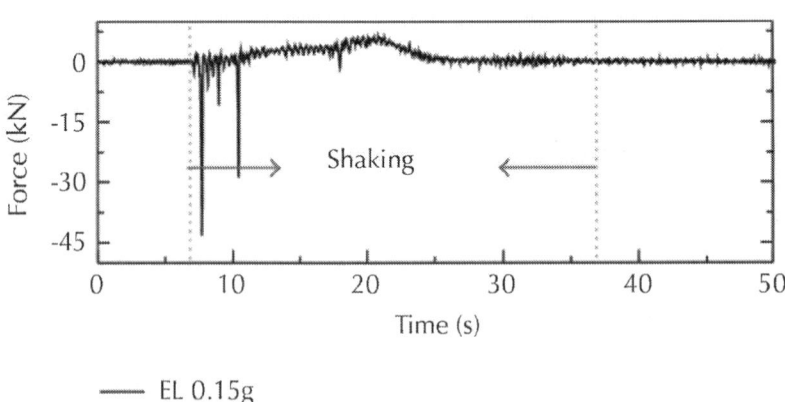

EL 0.15g

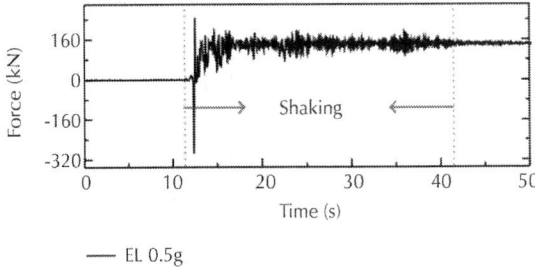

**Figure 10**: Increment of stabilizing force of pipe 2# under El-Centro wave in test 1.

Figure 11 gives the relationship between the stabilizing force and the excess pore pressure. The abscissa is the stabilizing force at the end of shaking for all tests, and the ordinate is the maximum value of P5. The stabilizing force shows a power function relationship with excess pore pressure. It can be seen that the stabilizing force is larger while the buried depth of the pipe is shallower at the same excess pore pressure ratio. It is probably because that the lateral constraint pressure of soil at shallow depth is smaller than that in deep place, so that the deflection deformation of shallower soil layer can be more intense which makes the pipe uplift more easily.

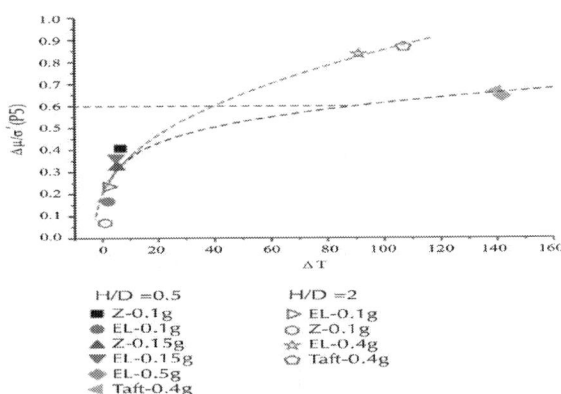

**Figure 11**: Relationship among increment of stabilizing force, Du/s' and excitation.

# ANALYSIS OF THE FORCE OF PIPE

The force components acting on pipes were investigated with the results obtained from centrifuge tests in this section. These components were adapted from static analysis to a dynamic condition where soil liquefaction occurs. Furthermore, the force analysis was validated by the comparison between the measured and calculated data.

## Force Analysis before Shaking

The force state before shaking is shown in Figure 12. And force equilibrium equation is expressed as follows:

$$T + W_s + \int_{-D/2}^{D/2} (u_1) dD \cdot L + W_p = \int_{-D/2}^{D/2} (u_2) dD \cdot L + N, \qquad (1)$$

Where L and $W_p$ are the length and weight of the pipe, respectively, which can be obtained according to scaling principle from the length and density of pipe in 1g condition. T is the stabilizing force of pipe 2# which can be measured by load cells (T of pipe 1# is 0). $W_s$ is the effective weight of overlying soil; N is the support force from the soil underlying the pipe; $\mu_1$ and $\mu_2$ are pore pressures around the pipe. The total force of these three parts can be calculated by the integral of the earth pressure difference between the upper and the bottom of pipe. It can be written as

$$P = \int_{-D/2}^{D/2} (u_2 - u_1) dD \cdot L + N - W_s, \qquad (2)$$

Where $P$ denoted the integral of earth pressure difference, which can be obtained by the interpolation according to the earth pressure transducers distributed on the pipes, defined as (3). Consider

$$P = \int_{-D/2}^{D/2} (e_2 - e_1) dD \cdot L, \qquad (3)$$

Where $e_1$ and $e_2$ denote the vertical component of linearized earth pressure distributed on the upper half and on the invert half of the pipe, respectively. The earth pressure is the total stress, which contains both effective stress and pore water pressure. The earth pressure mentioned below has the same meaning.

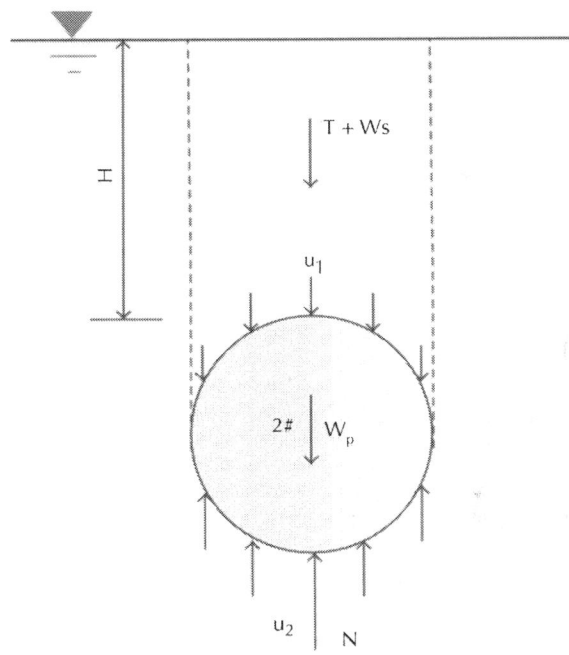

**Figure 12**: Schematic diagram of force of pipe 2# before shaking.

Equation (2) is checked based on the data measured before shaking, which proved to be reasonable with only a little bit difference as stresses around the pipe are estimated based on the interpolating method.

It should be noted that the integral of the pore pressure around the pipe under static state is buoyancy force, which is the static buoyancy force in the present study and can be calculated based on Archimedes principle as follows:

$$\int_{-D/2}^{D/2} (u_2 - u_1) dD \cdot L = \rho_w g V_{pipe.} \qquad (4)$$

# Force Analysis during Shaking For Pipe 2#

As pipe 2# was fixed to the rigid container, no displacement in the vertical direction occurred during vibration. And consequently the shear strength of soil could not excite. The force state of pipe 2# during vibration is shown in Figure 13. And force equilibrium equation is expressed in as follows:

$$T' + W_s' + \int_{-D/2}^{D/2} (u_1') dD \cdot L + W_p$$

$$= \int_{-D/2}^{D/2} (u_2') dD \cdot L + N', \qquad (5)$$

where the superscript sign (') represents the corresponding forces or stresses during shaking of pipe 1. Subtracting (1) from (5) gives

$$\Delta T = \int_{-D/2}^{D/2} (\Delta u_2 - \Delta u_1) dD \cdot L + \Delta N - \Delta W_s, \qquad (6)$$

Where DT is the increment of the stabilizing force which can be measured by the load cell. And are excess pore pressures around the

pipe. $\int_{-D/2}^{D/2} (\Delta u_2 - \Delta u_1) dD \cdot L$ is the integral of excess pore pressure during

soil liquefaction, labelled as $\Delta U \cdot \Delta W_s$ and $\Delta N$ are the increments of support force of soil underlying the pipe and the effective weight of soil overlying the pipe, respectively, due to the flowing deformation of soil around the pipe during vibration [22]. The three parts on the right-hand side of (6) can also be calculated by the integral of earth pressure differences around the pipe, marked as DP.

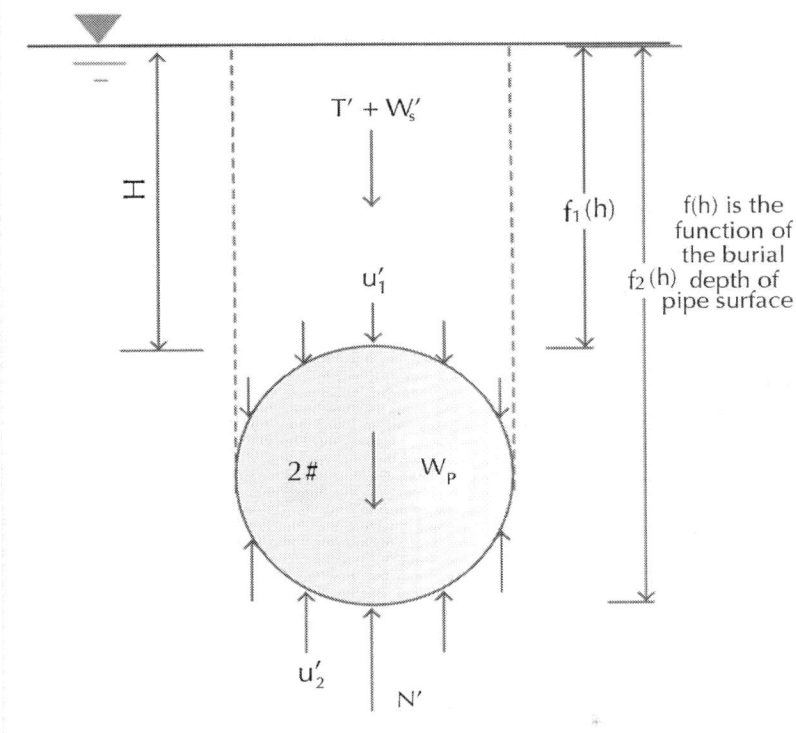

**Figure 13:** Schematic diagram of force of pipe 2# during shaking.

DP, DT, and DU of pipe 2# under 0.5 g excited by El-Centro wave are given in Figure 14. It is clear that the growth patterns of Dp and DT are almost the same. As the number of earth pressure transducers installed around the pipe is limited, the slight difference between DP and DT is reasonable. The integral of the excess pore pressure, DU, which is defined as the dynamic buoyancy force in this paper, is smaller than both DP and DU. Obviously, the uplifting behavior of pipe is not only affected by the build-up of the excess pore pressure, but also by the variations of the effective weight of the overlying soil and the support force of the underlying soil.

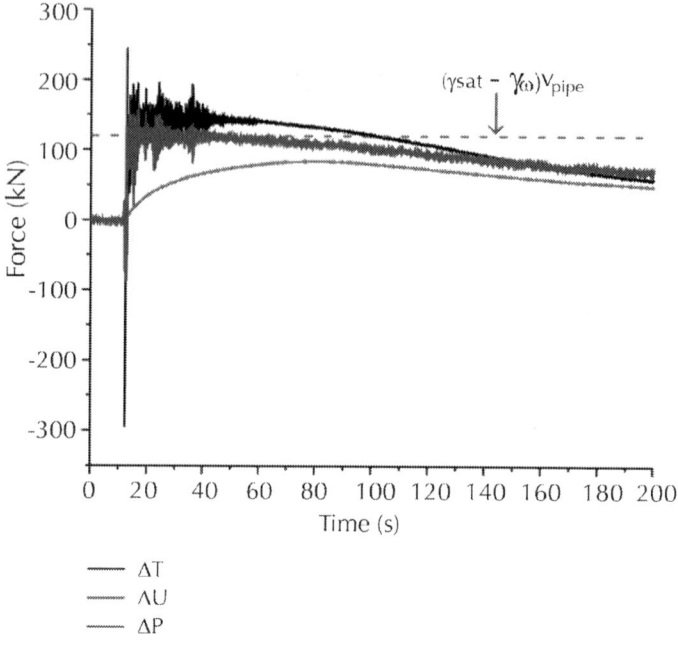

**Figure 14:** Relationship among and under 0.5 g El-Centro wave in test 1.

The value of $(\gamma_{sat} - \gamma_w) \cdot V_{pipe}$, which is commonly adopted by other researchers to calculate the buoyancy force, is given in Figure 14. When the soil is fully liquefied, the excess pore pressure then reaches the value of the initial effective vertical earth pressure (the pore pressure at the crown surface of the pipe is $u_1' = \gamma_{sat} f(h_1)$, and the value on the bottom of the pipe is $u_2' = \gamma_{sat} f(h_2)$). Integrating the difference between $u_1$ and $u_2$ gives

$$\int_{-D/2}^{D/2} \left( u_2' - u_1' \right) dD \cdot L$$

$$= \gamma_{sat} \cdot \int_{-D/2}^{D/2} \left[ f(h_2) - f(h_1) \right] dD \cdot L = \gamma_{sat} V_{pipe}$$

$$(7)$$

It indicates that the dynamic buoyancy force DU is equal to $(\gamma_{sat} - \gamma_w) \cdot V_{pipe}$ for fully liquefied ground. The value of dynamic buoyancy force will be overestimated for incompletely liquefied soil by (7).

# Force Analysis during Shaking for Pipe 1#

The force state of pipe 1# during uplifting is shown in Figure 15. And force equilibrium equation is expressed as follows:

$$F_s + W_s^{''} + \int_{-D/2}^{D/2} u_1^{''} dD \cdot L + W_p$$

$$= \int_{-D/2}^{D/2} u_2^{''} dD \cdot L + N^{''} + m_p a, \tag{8}$$

Where a represents the uplift acceleration of pipe; $F_s$ represents the frictional resistance from the overlying soil, which varies with the degree of soil liquefaction and reduces to 0 if the soil is fully liquefied. The physical meanings of $u_1^{''}, u_2^{''}, W_s^{''},$ and $N^{''}$ are the same as $u_1^{'}, u_2^{'}, W_s^{'},$ and $N^{'}$, except the superscript sign ('') represents the corresponding forces or stresses during vibration of pipe 1#. Subtracting (1) from (8) gives

$$m_p a = F_s + \Delta^* W_s - \int_{-D/2}^{D/2} \left( \Delta^* u_2 - \Delta^* u_1 \right) dD - \Delta^* N, \tag{9}$$

Where $\Delta^* u_1$ and $\Delta^* u_2$ are excess pore pressures around the pipe; $\int_{-D/2}^{D/2} \left( u_2^{'} - u_1^{'} \right) dD \cdot L$ is dynamic buoyancy force during soil liquefaction, labelled as $\Delta^* U \cdot \Delta^* W_s$ and $\Delta^* N$ are the increments of support force of soil underlying pipe 1# and the effective weight of soil overlying pipe 1#, respectively, due to the excess pore pressure variation induced flowing deformation of soil around the pipe during vibration. As the different motion patterns of pipe 1# and pipe 2#, the stresses measured by the transducers around them are different. The total force of the three parts can also be calculated by the integral of earth pressure around pipe 1#. All the forces in (9) can be calculated by the measured data except Fs.

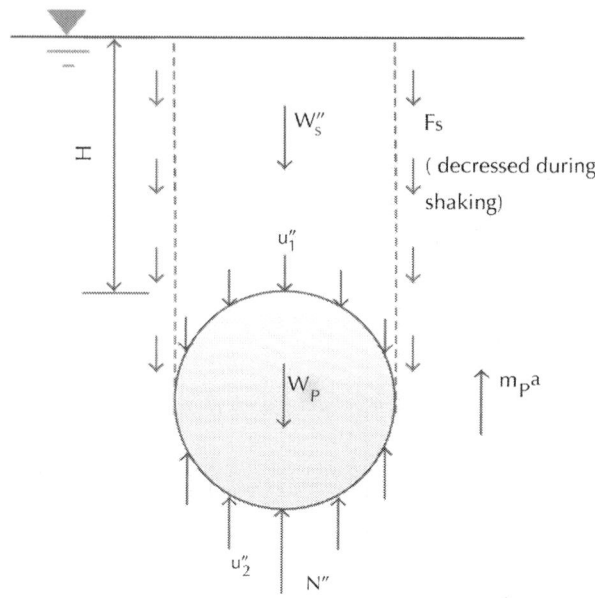

**Figure 15**: Schematic diagram of force of pipe 1# during shaking.

The frictional force $F_s$ from the shear plane is estimated using the equation introduced by DNV (2007) [27] as follows:

$$F_S = f_p \left[ \frac{D}{H} \times \left( \frac{H}{D} + 0.5 \right)^2 \right] \sigma'_H DL,$$

(10)

where $\sigma'_H$ refers to the effective vertical stress of the overlying soil. Before vibration $\sigma'_{H0} = \gamma' H \cdot f_p$ is a parameter related to the soil property ranging from 0.4 to 0.6 for medium dense sand.

Given the relationship between the degree of liquefaction and the frictional contact between the soil grains, vertical effective stress declines linearly with the increase of excess pore pressure. Consider

$$\sigma_H' = \sigma_{H0}' - \Delta u. \tag{11}$$

Substituting (10) and (11) into (9), the uplift acceleration (a) can be obtained. And acceleration time-history is shown in Figure 16(a). The concept of the Newmark's method which is used to predict earthquake-induced permanent deformation is adopted in the case of a floating pipe [28]. The angle of the sliding surface is considered to be vertical rather than inclined. And the rigid pipe can be treated as sliding block of the Newmark's model with vertical movement.

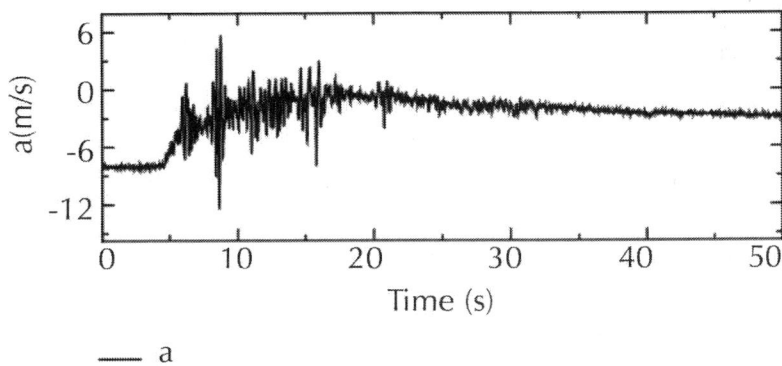

Figure 16: Acceleration, velocity and displacement of pipe 1# under 0.4 g Taft wave in test 1.

As the behavior of pipe in liquefied soil is extremely complicated, a few assumptions are made in the following to calculate the displacement of pipe. The uplifting of the pipe 1# occurs as soon as the value of the uplift acceleration is greater than zero. Zero is deemed as the yield acceleration. The movement of pipe occurs when its acceleration exceeds the yield acceleration, which is in accordance with Newmark's analysis. The excess in acceleration above yield acceleration is termed as effective vertical acceleration (). And effective vertical acceleration time-history is illustrated in Figure 16(b). In addition, pipe is incapable of sinking considering the bearing capacity of the underlying soil and the flowing of soil from the top or side to the bottom of the pipe. As pipe 1# is in the static status before shaking, the initial value of acceleration, velocity, and displacement should be zero.

Based on the assumptions above, the accumulated uplifting displacement can be obtained by integrating the effective vertical acceleration $a_{eff}$ twice. The uplifting displacement time history of the pipe 1# calculated under 0.4 g Taft wave in test 1 is shown in Figure 16(d). The predicted displacements are slightly larger than the experimental ones observed in Figure 16 and fluctuate around the experimental ones in other cases. The difference between the measured and the predicted displacements is lower than 20 mm, which is deemed to be acceptable as the number of transducers installed around the pipe is limited. And the motion patterns of them are almost the same. Therefore, the proposed approach to estimating the stabilizing force around pipe is reasonable.

# CONCLUSIONS

The uplifting behavior of shallow buried pipe in liquefied field was investigated through dynamic centrifuge model tests in the present study, and the main conclusions of the research are summarized as follows.

1.  Although the uplifting phenomenon of pipelines in the liquefied soils always happens during the seismic vibration, the observation in our tests shows the begin and end time point of uplifting is not directly related to the seismic motion. The uplifting is highly dependent on the buildup of the excess pore pressure. Moreover, the quantitative relationship between the uplifting behavior and the generation of the excess pore pressure needs further studies.

2.  The uplifting movement of pipe is the combination effects of multiple forces. During seismic vibration, excess pore pressure generates and soil around the pipeline gradually flow in an oval-like trace, which causes both the variation of effective weight of overlying soil and supporting force of soil underlying the pipeline, as well as the shear resistance from shear planes that varies with the degree of liquefaction. As a result, the equilibrium of pipeline during shaking is broken and the pipe consequently uplifts. However, in most existing research, the variations of the overlying soil weight and the supporting force of the underlying soil are ignored.

3.  For incompletely liquefied field, the buoyancy force is overestimated by multiplying the saturated unit weight of soil and pipeline volume.

# CONFLICT OF INTERESTS

The authors declare that there is no conflict of interests regarding the publication of this paper.

# ACKNOWLEDGMENTS

This research work was supported by National Natural Science Foundation of China (nos. 51178427 and 51278451) and Key Innovation team support project of Zhejiang Province (2009R50050).

# REFERENCES

1.  M. Kitaura and M. Miyajima, "Quantitative evaluation of damage to buried pipelines induced by soil liquefaction," in Proceedings of the 9th World Conference on Earthquake Engineering, pp. 11–16, Tokyo, Japan, August 1988.

2.  S. Yasuda and H. Kiku, "Uplift of sewage manholes and pipes during the 2004 Niigataken-Chuetsu Earthquake," Soils and Foundations, vol. 46, no. 6, pp. 885–894, 2006

3.  W. J. Hall and T. D. O'Rourke, "Seismic behavior and vulnerability of pipelines," Lifeline Earthquake Engineering, ASCE, pp. 761–773, 1991.

4.  T. D. O'Rourke and P. A. Lane, Liquefaction Hazards and Their Effects on Buried Pipelines, National Center for Earthquake Engineering Research, 1989.

5.  T. D. O'Rourke, T. E. Gowdy, H. E. Strwart, and J. W. Pease, "Lifeline and geotechnical aspects of the 1898 Loma Prieta Earthquake," in Proceedings of the 2nd International Conference on Recent Advances in Geotechnical Earthquake Engineering and Soil Dynamics, pp. 1601–1612, University of Missouri-Rolla, Rolla, Mo, USA, 1991.

6.  Y. Mohri, M. Yasunaka, and S. Tani, "Damage to buried pipeline due to liquefaction induced performance at the ground by the Hokkaido-Nansei-Oki Earthquake in 1993," in Proceedings of the 1st International Conference on Earthquake Geotechnical Engineering, pp. 31–36, Tokyo, Japan, 1995.

7.  O. Kiyomiya and K. Minami, "Evaluation of stresses on submarine pipelines in liquefied seabed," inProceedings of the 9th World Conference on Earthquake Engineering, pp. 91–96, Tokyo-Kyoto, Japan, August 1988.

8.  X. Zhu, S. Xue, X. Tong, and X. Sun, "Uplift response of large-diameter buried pipeline in liquefiable soil using pipe-soil coupling model," in Proceedings of the ASCE-International Conference on Pipelines and Trenchless Technology (ICPTT '11), pp. 1790–1801, Beijing, China, October 2011. View at Publisher ·View at Google Scholar · View at Scopus

9.  M. Kitaura, M. Miyajima, and H. Suzuki, "Response analysis of buried pipelines considering rise of ground water table in liquefaction processes," Japan Society of Civil Engineers, vol. 4, no. 1, pp. 147–154, 1987.

10. H. I. Ling, Y. Mohri, T. Kawabata, H. Liu, C. Burke, and L. Sun, "Centrifugal modeling of seismic behavior of large-diameter pipe in liquefiable soil," Journal of Geotechnical and Geoenvironmental Engineering, vol. 129, no. 12, pp. 1092–1101, 2003.

11. Z. Xia, G. Ye, J. Wang, B. Ye, and F. Zhang, "Numerical analysis on the influence of thickness of liquefiable soil on seismic response of underground structure," Journal of Shanghai Jiaotong University, vol. 15, no. 3, pp. 279–284, 2010 (Chinese).

12. R. Saeedzadeh and N. Hataf, "Uplift response of buried pipelines in saturated sand deposit under earthquake loading," Soil Dynamics and Earthquake Engineering, vol. 31, no. 10, pp. 1378–1384, 2011.

13. D. G. Zou, X. J. Kong, H. I. Ling, and T. Zhu, "Experimental study on the uplift behavior of pipeline in saturated sand foundation and earthquake resistant measures during an earthquake," Chinese Journal of Geotechnical Engineering, vol. 24, no. 3, pp. 323–326, 2002 (Chinese).

14. S. C. Chian and S. P. G. Madabhushi, "Displacement of tunnels in lquefied sand deposits," inProceedings of the 8th International

Conference on Urban Earthquake Engineering, pp. 517–522, Tokyo, Japan, 2011.

15.  L. Sun, Centrifuge Modeling and Finite Element Analysis of Pipeline Buried in Liquefiable Soil, Columbia University, 2001.

16.  N. Nishio, "Mechanism of projection of sewerage manholes above ground due to soil liquefaction,"Japan Society of Civil Engineers, vol. 11, no. 3, pp. 145–148, 1994.

17.  K. Sekiguchi, S. Matsuda, and H. Adachi, "Numerical study on the effectiveness of stabilizing techniques of offshore pipelines against liquefaction," in Proceedings of the 11th World Conference on Earthquake Engineering, 1996.

18.  J. Q. Lin, Z. H. Li, and M. Y. Hu, "Study on the floatation response of buried pipelines due to soil liquefaction," Journal of Earthquake Engineering and Engineering Vibration, vol. 24, no. 3, pp. 120–123, 2004 (Chinese).

19.  S. C. Chian and K. Tokimatsu, "Floatation of underground structures during the Mw9.0 T hoku earthquake of 11th March 2011," in Proceedings of the 15th World Conference on Earthquake Engineering, Lisboa, Portugal, 2012.

20.  M. Y. Chen, C. Han, D. S. Ling, L. G. Kong, and Y. G. Zhou, "Development of geotechnical centrifuge ZJU400 and performance assessment of its shaking table system," Chinese Journal of Geotechnical Engineering, vol. 33, no. 12, pp. 1887–1894, 2011 (Chinese).

21.  R. S. Steedman and S. P. G. Madabhushi, "Wave propagation in sand medium," in Proceedings of the International Conference on Seismic Zonation, Stanford, California, 1991.

22.  S. C. Chian, Floatation of Underground Structures in Liquefiable Soil, University of Cambridge, 2012.

23.  L. G. Kong, J. Y. Fan, R. P. Chen, and Y. M. Chen, "Pile-soil-pile interaction between two piles moving along different directions," Chinese Journal of Geotechnical Engineering, vol. 33, no. 12, pp. 1887–1894, 2011 (Chinese).

24.  Y. Guo, M. T. Luan, C. S. Xu, and Y. He, "Effect of variation of principal stress orientation on undrained dynamic strength behavior of loose sand," Chinese Journal of Geotechnical Engineering, vol. 25, no. 6, pp. 666–670, 2003 (Chinese).

25.  D. Yang, E. Naesgaard, P. M. Byrne, K. Adalier, and T. Abdoun, "Numerical model verification and calibration of George Massey Tunnel using centrifuge models," Canadian Geotechnical Journal, vol. 41, no. 5, pp. 921–942, 2004.

26.  T. Yi, H. Li, and M. Gu, "Optimal sensor placement for structural health monitoring based on multiple optimization strategies," The Structural Design of Tall and Special Buildings, vol. 20, no. 7, pp. 881–900, 2011.

27.  J. Wang, S. K. Haigh, G. Forrest, and N. I. Thusyanthan, "Mobilization distance for upheaval buckling of shallowly buried pipelines," Journal of Pipeline Systems Engineering and Practice, vol. 3, no. 4, pp. 106–114, 2012.

28.  N. M. Newmark, "Effect of earthquakes on dams and embankments," Milestones in Soil Mechanics, pp. 109–129, 1965.

# 10

# Numerical Simulation on Effects of Electromagnetic Force on The Centrifugal Casting Process of High Speed Steel Roll

Minghu Yuan, Leilei Cao, Yaozeng Xu, and Xuding Song

Key Laboratory of Road Construction Technology & Equipment of MOE, Chang'an University, Xi'an, China.

## ABSTRACT

A three-dimensional mathematical and physical model coupling with the heat transfer and the flow of molten metal in the centrifugal casting of the high speed steel roll was established by using CFD software FLUENT. It can be used to analyze the distribution of the temperature filed and the flow filed in the centrifugal casting under the gravity, the electromagnetic stirring force and the centrifugal force. Some experiments were carried out to verify the above analysis results. The effects of the electromagnetic force on the centrifugal casting process are discussed. The results showed that under the 0.15 T electromagnetic field intensity, both the absolute pressure of metal flow to mold wall and the metal flow velocity on the same location have some differences between the electromagnetic centrifugal casting and the centrifugal casting. Numerical results for understanding the electromagnetic stirring of the centrifugal casting process have a guiding significance.

# INTRODUCTION

High speed steel roll has been widely used in the rolling production because of its excellent wear resistance and the better performance than the traditional steel roll [1,2]. In the process of the electromagnetic centrifugal casting, the molten metal flow has a great influence on the quality and the performance of the roll. Since the centrifugal casting is under the complicated force situation and under the high speed, the high temperature and the opaque environment, it is difficult to know the moving and the filling rule of the molten metal. Therefore, it is necessary to analyze the flow field of the molten metal in the electromagnetic centrifugal casting process. Researches about the electromagnetic centrifugal casting mainly focus on the as-cast microstructure. There are less researches on the moving and the filling rule of the molten metal in the electromagnetic centrifugal casting process. Numerical method provides a new way to solve this problem. But the existing numerical methods are based on many simplifications and assumptions. For example, the flowing of the molten metal and the heat transfer for it are analyzed independently; suppose that the gravity can be neglected and the molten metal fills cast instantneously [3-5]. These simplifications and assumptions make the large gap between numerical results and true values. In this paper, a three-dimensional model coupling of the heat transfer and the flow of the molten metal in the centrifugal casting of the high speed steel roll is established by the finite volume method. The molten metal flow pattern has been obtained. The effects of the rotational speed and the electromagnetic force on the centrifugal casting process have been discussed. The results can provide the theoretical evidence for setting the parameter and improving the cast quality.

# MATHEMATICAL AND PHYSICAL MODEL ESTABLISHED

In the centrifugal casting process, molten metal is under gravity, centrifugal force and also, electromagnetic force. When molten metal in the centrifugal mold rotated at high speed with a stationary magnetic field, the electromagnetic force compels molten metal to flow in the opposite direction of the rotating mold, resulting in electromagnetic stirring, which can enhance solidification structure and improve casting quality [6-8]. Under gravity, the molten metal poured into rotating mold taking the form of parabola. Then, the friction force between molten metal and mold, the viscous force of molten metal and the centrifugal force generated by the high speed revolution make molten metal into hollow roll. The moving of molten metal is determined by the force, and also, the temperature. Therefore, it is necessary to analyze the behavior of molten metal by coupling temperature field, flow field and magnetic field. The mass equation, momentum equation, energy equation and electromagnetic force equation which describe the moving of the molten metal are listed as below, respectively.

Mass equation :

$$\frac{\partial \rho}{\partial t} + \nabla(\rho v) = \frac{\partial \rho}{\partial t} + \rho \nabla v + v \nabla \rho = 0 \tag{1}$$

Momentum equation:

$$\rho \nabla v^2 + v \nabla \rho = -\nabla \rho + \mu \nabla^2 v + \nabla \mu . \nabla v + F \tag{2}$$

Energy equation:

$$\rho C(T) \frac{dT}{dt} = \lambda(T) \nabla^2 T + \nabla \lambda(T) . \nabla T \tag{3}$$

where   Ñ is Hamiltonian operator, $\nabla = \dfrac{\partial}{\partial x}i + \dfrac{\partial}{\partial y}j + \dfrac{\partial}{\partial z}k$; Ñ² is

Laplace operator, $\nabla^2 = \dfrac{\partial^2 ()}{\partial x^2} + \dfrac{\partial^2 ()}{\partial y^2} + \dfrac{\partial^2 ()}{\partial z^2}$; $\rho$, $v$, P, F, C, $\lambda$ and T are the density, velocity vector, pressure, body force, specific heat, thermal conductivity and temperature, respectively.

Electromagnetic force can be calculated by Maxwell equation:

$$F_{em} = \sigma_e \omega r B \cos \omega t \tag{4}$$

Where $F_{em}$, $\sigma_e$, r, t, $\omega$B present electromagnetic force; conductivity; radius; time; electromagnetic induction intensity; the angular velocity of molten metal in the mold respectively. B = 015T is employed in this paper.

The physical model is built by the pre-processing software Gambit. The principle of electromagnetic centrifugal cast is shown in Figure 1. x, y, and z present horizontal radial, vertical radial and axial direction, respectively. The origin of coordinates is set on the center of rotation. The physical mesh model is shown in Figure 2. In this model, 73,906 hexahedral elements are employed to mesh solution domain.

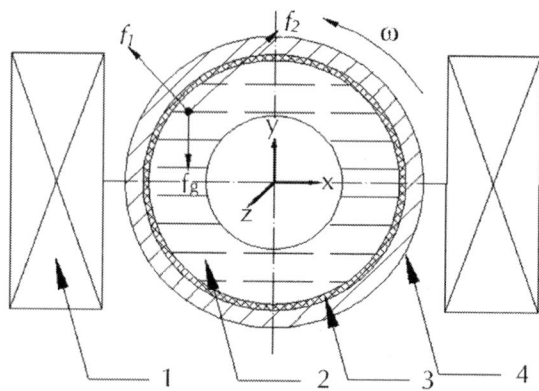

**Figure 1.** The principle schematic diagram of electromagnetic centrifugal casting (1-electromagnet; 2-molten metal; 3-coating; 4-mold; $f_1$-centrifugal force; $f_2$-electromagnetic force; $f_g$-gravity; -angular velocity of mold).

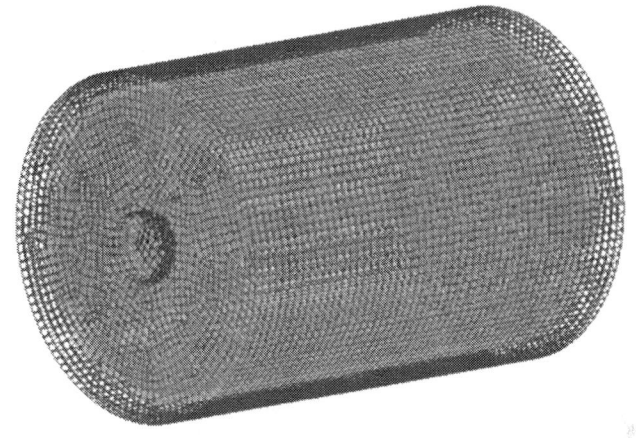

**Figure 2**. Physical mesh.

The material of mold is stainless steel and its inner surface is covered by coating which made of water glass and hardening quartz sand. The external diameter and inner diameter of the mold are 200 mm and 180 mm respectively. Its depth is 250 mm. The rotational speed can be set as 800 r/min, 1100 r/min or 1600 r/min. The initial temperature of the mold is 300 K and the pouring temperature is 1833 K.

VOF model is chosen for multiphase flow to track free interface between molten metal and air. The solver which based on pressure and the first order implicit time scheme are employed in VOF model. Reynolds stress model is chosen as turbulence model because it can simulate high intensity eddy and anisotropy of turbulence very well. Solidification and melting model are used to simulate phase change process.

The magnetic field of the air gap of electromagnet is not uniform in reality. When the air gap is small, it can be considered as uniform. In the numerical analysis, the magnetic field is considered as uniform and its intensity value should be put into the software as input. The material properties of stainless steel are as follows, thermal conductivity is $57.8 W.(m.K)^{-1}$; specific heat is $481\,481 J.(kg.K)^{-1}$ and density is $7800 kg.m^{-3}$. For quartz sand, thermal conductivity is $1.76 W.(m.K)^{-1}$; specific heat is $2570 J.(kg.K)^{-1}$ and density is $1830 kg.m^{-3}$. Thermal properties of high speed steel are shown in Table 1.

**Table 1:** Thermal physical properties of high speed steel

| Thermal conductivity /W (m K)<sup>-1</sup> | Specific heat | Density | Dynamic viscosity | Liquidus temperature | TL/K Solidus temperature TS / K |
|---|---|---|---|---|---|
| 25.5 | 500 | 7700 | 0.0092 | 1695 | 1539 |

# NUMERICAL SIMULATION RESULTS AND ANALYSIS

## The Influence of Rotational Speed on Centrifugal Casting Process

In this example, centrifugal casting process is considered. The size of the sprue is $\Phi$ 40 mm; the pouring velocity is 0.15 m/s; the thickness of the roll is 30 mm. It is assumed that molten metal is incompressible, Newtonian fluid and its material properties are not changing with temperature. In the calculation, time step is chosen as 0.005 s - 0.01 s to adapt different Courant value. This simulation can simulate the whole process of centrifugal casting for a period of 22.8 second.

In general, rotational speed of centrifugal casting process has a critical value. When rotational speed researches the critical value, the high quality roll which has predetermined shape can be obtained. When the rotational speed is much lower than the critical value, eccentricity may occur in roll because of centrifugal force deficiency, and make the deformation unsuccessful. While, when the rotational speed is much higher than the critical value, large tensile stress may occur and longitudinal crack is developed, which makes segregation even worse. Figure 3 show molten metal volume fraction distribution in the period of 22.8 s under rotational speed 1100 r/min.

It can be seen from Figure 3 that at the initial stage when molten metal pouring into the mold, taylor vortex flow happens. The eddy pitch decreases with increasing rotational speed. Also, it can be seen from Figure 3(f) and Figure 4 that distributions of molten metal volume fraction have eccentric phenomena at the end of the pouring process.

Moreover, the eccentric phenomena get improved when rotational speed increasing. Figure 5 is the picture of high speed steel roll after machining, the eccentric phenomena can also be seen clearly form Figure 5. Comparing the results for different rotational speed (800 r/min, 1100 r/min, 1600 r/min), we can see the smoothest molten filling is got with a rotational speed of 1100 r/min and the inner surface of the roll is more uniform at 1100 r/min, which has significant influence on the cast quality.

# The Influence of Electromagnetic Force on Absolute Pressure in the Mold

In the electromagnetic centrifugal casting process, molten metal is under gravity, centrifugal force and also, electromagnetic stirring force generated by electromagnetic induction. The direction of the tangential component of electromagnetic force is opposite to the motion direction of the molten metal, which makes relative movement between molten metal. This relative movement may change flow filed, solute redistribution of molten metal and metal solidification process, and then, change microstructure and performance of the cast.

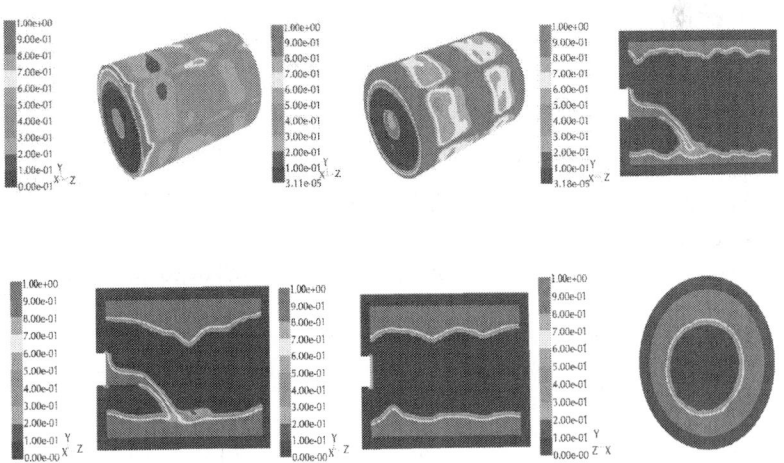

**Figure 3.** Molten metal volume fraction distribution under rotational speed of 1100 r/min. (a) 2 s; (b) 4 s; (c) 10s; (d) 16.8 s; (e) 22.8 s; (f) the cross-section from sprue 120 mm at 22.8 s.

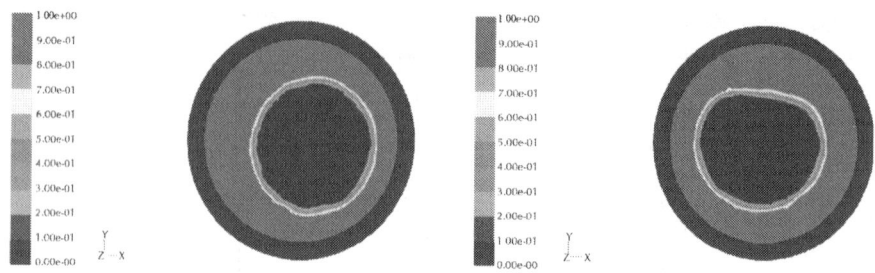

**Figure 4:** Molten metal volume fraction distribution of the cross-section from sprue 120 mm at 22.8 s. (a) under 800 r/mm; (b) under 800 r/mm.

**Figure 5.** The picture of high speed steel roll after machining outer surface.

The effects of electromagnetic filed on casting process can be observed directly by the change of absolute pressure to the mold. Electromagnetic filed makes the centrifugal force decreased to some extent. Comparing the results under different conditions, which have electromagnetic field applied or not, it can be seen form Figures 6 and 7 that differences exist on the absolute pressure on the same location. The difference is small since the electromagnetic force is relative small compared with centrifugal force, and the difference ranges from 25 Pa to 2418 Pa.

## The Influence of Electromagnetic Force on Metal Flow Velocity

The flowing state of the molten metal is basically identical with centrifugal casting. On the transverse interface of the mold, the electromagnetic force compels molten metal to flow in the opposite direction of the rotating mold, so the velocity vector of the molten metal decreases when applying electromagnetic field.

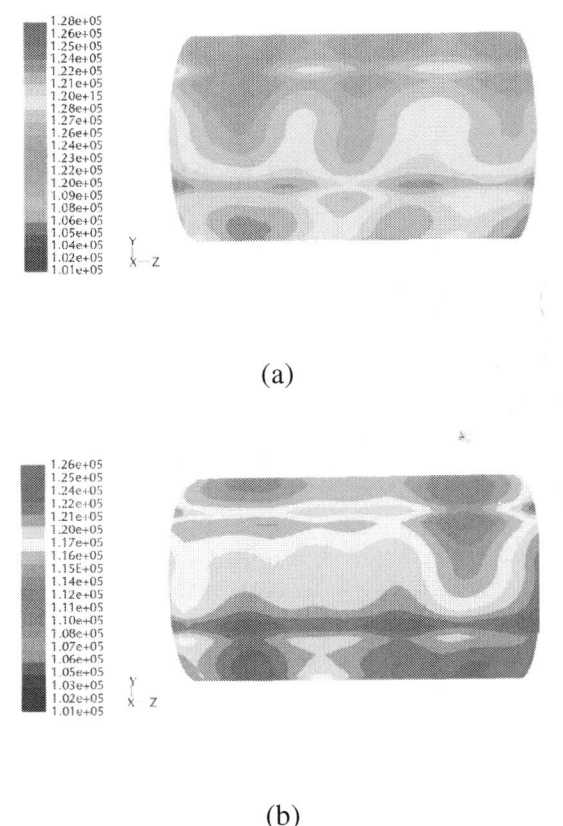

(a)

(b)

**Figure 6:** The three dimensional picture of absolute pressure vary on the electromagnetic centrifugal casting and the centrifugal casting.

Figure 8 is the comparison of resultant velocity between centrifugal casting and electromagnetic centrifugal casting from sprue 120 mm under rotational speed 1100 r/m. Since the molten metal is flowing fast, it is in turbulent state. The adjacent flow mixes up and produce velocity component which is perpendicular to horizontal cross-section. In the whole, the velocity of molten metal may decrease after applying magnetic field. Comparing the resultant velocity with that of the ordinary centrifugal casting, the maximal difference is 0.104 m/s and the minimal difference is 0.064 m/s. This reflects the electromagnetic stirring function.

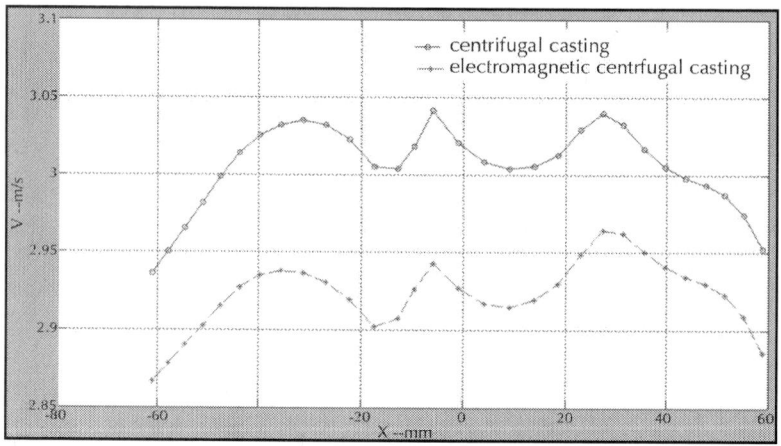

**Figure 7.** The comparison curve of absolute pressure vary on the electromagnetic centrifugal and the centrifugal casting casting.

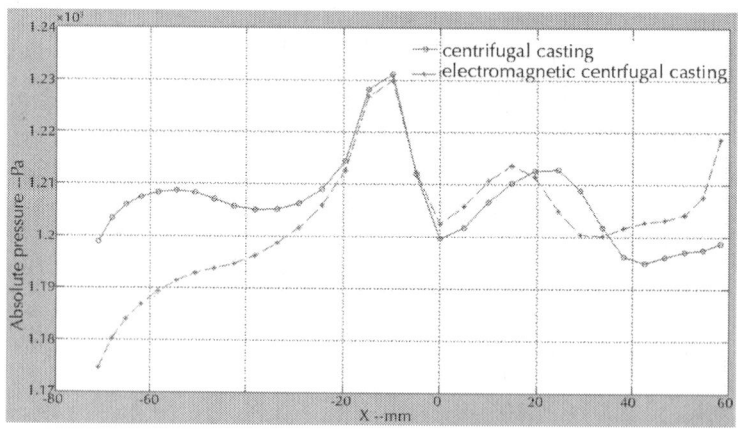

**Figure 8:** The comparison curve of resultant velocity between centrifugal and electromagnetic centrifugal casting.

# CONCLUSIONS

In the centrifugal casting process, since the molten metal is under the gravity, the centrifugal force and the electromagnetic force, the eccentricity may happen to the roll. The numerical results agree well with the experiment results.

The high quality roll which has the predetermined shape can be gained when the casting operated at the critical rotational speed. In this paper, the smoothest molten filling can be obtained with a rotational speed of 1100 r/min.

The simulation is conducted under the 0.15 T electromagnetic field intensity and the results show that both the absolute pressure of the metal flow to mold wall and the metal flow rate on the same location have some differences between the electromagnetic centrifugal casting and the centrifugal casting. This reflects the electromagnetic stirring function.

# ACKNOWLEDGEMENTS

The authors would like to thank the Project supported by the Special Fund for Basic Scientific Research of Central Colleges, Chang'an University (CHD2013G1250133, CHD2013G3252006, and CHD2013G2251007).

# REFERENCES

1.   Y.-L. He, Y.-S. Yang and Z.-L. Hu, "Finite Element Simulation of the Melt Flow and Heat Transfer in Electromagnetic Centrifugal Casting," Foundry, Vol. 49, No. 8, 2000, pp. 473-477

2.   T.-M. Zhang, Y.-T. An, X.-D. Song, et al., "Influence of Magnetic Flux Density on Microstructure and Properties of High Carbon High Speed Steel Made by Electromagnetic Centrifugal Casting," Journal of Iron and Steel Reseach, Vol. 24, No. 5, 2012, pp. 35-40.

3.   B. Chen, M. GAO, Y.-C. Ma, et al., "Numerical Simulation on the Electromagnetic Centrifugal Casting of Tube Blank," Journal of Iron and Steel Reseach, Vol. 19, No. 6, 2005, pp. 631-638.

4.   K. S. Keerthiprasad and M. S. Murali, "Numerical Simulation and Cold Modeling Experiments on Centrifugal Casting," Metallurgical and Materials Transactions, BProcess Metallurgical and Materials Processing Science, Vol. 142, No. 1, 2011, pp. 144-155.

5.   L.-G. Liu, Q.-X. Yang, Q. Li, et al., "Simulation of Temperature and Stress Fields during the Electromagnetic Centrifugal Casting Process," Journal of University of Science and Technology Beijing, Vol. 28, Suppl. 1, 2006, pp. 523-525.

6.   H.-G. Fu and J.-D. Xing, "Manufacturing Technology of High Speed Steel Roll," Metallurgical Industry Press, Beijing, 2007.

7.   Q.-J. Wang, "Study on the Microstructures and Properties of High Speed Steel Roll by Electromagnetic centrifugal casting," Chang'an University, Chang'an, 2010.

8.   W.-Q. Zhang, "The Principle and Technology of Metal Electromagnetic Solidification," Metallurgical Industry Press, Beijing, 2004.

# Citations

## CHAPTER 1

Loli M., Knappett J. A., Brown M. J., Anastasopoulos I., and Gazetas G. (2014), Centrifuge modeling of Rocking isolated Inelastic RC Bridge Piers, Earthquake Engineering & Structural Dynamics, 30, pages 2341–2359, doi: 10.1002/eqe.2451

## CHAPTER 2

S W Jacobsz, Centrifuge Modelling of a Soil Nail Retaining Wall, ISSN 1021-2019.

# CHAPTER 3

Heon-Joon Park and Dong-Soo Kim, Centrifuge Modelling for Evaluation of Seismic Behaviour of Stone Masonry Structure, doi:10.1016/j.soildyn.2013.06.010.

# CHAPTER 4

Carlo DIETL and Hemin KOY, Formation of Tabular Plutons – Results and Implications of Centrifuge Modelling, DOI: 10.3190/jgeosci.032.

# CHAPTER 5

Amr Farouk Elhakim, Mohamed Abd Allah El Khouly, and Ramy Awad, Three Dimensional Modeling of Laterally Loaded Pile Groups Resting in Sand, doi:10.1016/j.hbrcj.2014.08.002.

# CHAPTER 6

Ioanna Ioannou, John Douglas, and Tiziana Rossetto, Assessing the Impact of Ground-Motion Variability and Uncertainty on Empirical Fragility Curves, doi:10.1016/j.soildyn.2014.10.024.

# CHAPTER 7

Liping Wang and Ga Zhang, Pile-Reinforcement Behavior of Cohesive Soil Slopes: Numerical Modeling and Centrifuge Testing, Journal of Applied Mathematics, vol. 2013, Article ID 134124, 15 pages, 2013. doi:10.1155/2013/134124.

# CHAPTER 8

Charles W. W. Ng, the State-of-The-Art Centrifuge Modelling of Geotechnical Problems at Hkust, doi: 10.1631/jzus.A1300217.

# CHAPTER 9

Bo Huang, Jingwen Liu, Peng Lin, and Daosheng Ling, Uplifting Behavior of Shallow Buried Pipe in Liquefiable Soil by Dynamic Centrifuge Test, The Scientific World Journal, vol. 2014, Article ID 838546, 15 pages, 2014. doi:10.1155/2014/838546.

# CHAPTER 10

M. Yuan, L. Cao, Y. Xu, and X. Song, Numerical Simulation on Effects of Electromagnetic Force on the Centrifugal Casting Process of High Speed Steel Roll, Modeling and Numerical Simulation of Material Science, Vol. 4 No. 1, 2014, pp. 20-24. doi: 10.4236/mnsms.2014.41004.

# Index